DATE DUE

MAR 2 1 2000		
APR 1 7 2000		
5-8-00		
GAYLORD		PRINTED IN U.S.A.

Animal Cognition

JACQUES VAUCLAIR

Animal Cognition

*An Introduction to Modern
Comparative Psychology*

Harvard University Press
Cambridge, Massachusetts
London, England 1996

Library of Congress Cataloging-in-Publication Data

Vauclair, Jacques.
   Animal cognition : an introduction to modern comparative psychology / Jacques Vauclair.
      p.   cm.
   "Portions of this volume are based on my book, L'intelligence de l'animal, which was
published in French by Editions du Seuil (Paris, 1992)"—Pref.
   Includes bibliographical references and index.
   ISBN 0-674-03703-0 (alk. paper)
   1. Cognition in animals. 2. Animal psychology. 3. Animal intelligence. I. Vauclair, Jacques.
   Intelligence de l'animal. II. Title.
QL785.V335 1996
591.51—dc20
95-46951

To Sébastien

Contents

Preface   ix

1. Origins and Development of the Study of Animal Cognition   1

   The Darwinian Heritage and Nineteenth-Century Psychology   *1*

   The Behaviorist Break   *3*

   The Emergence of the Cognitive Approach   *4*

   The Modern Concepts of Representation and Memory   *7*

   The Study of Representation in Animals   *8*

   Problems Posed by the Study of Cognition in Animals   *10*

2. Laboratory Methods for Assessing Representation in Animals   12

   Learning Sets   *12*

   Mastery of Relations between Stimuli   *13*

   Category Formation   *14*

   Serial Learning as Evidence of Nonverbal Thought   *16*

   Mental Images in Animals   *20*

   Summary and Current Debate   *27*

3. Piagetian Studies in Animal Psychology   29

   Developmental Psychology and Comparative Psychology   *29*

   The Development of Intelligence   *31*

   Sensorimotor Activities in Animals   *35*

   "Concrete Operations" in Animals   *42*

   Summary and Current Debate   *50*

4. Tool Use and Spatial and Temporal Representations   53

Tool Use   53

Spatial Representations   62

Temporal Representations   80

Summary and Current Debate   81

5. Social Cognition   84

Experimental Methods for the Study of Social Cognition   86

Social Cognition in Monkeys   87

Social versus Nonsocial Cognition   92

Suggestions for Future Research   95

Summary and Current Debate   96

6. Animal Communication and Human Language   99

Comparisons of Animal and Human Communication   101

Language-Trained Animals   106

Differences in the Use of Signs by Apes and Children   115

*Pre-Linguistic Communication in Human Infants and
Chimpanzee Infants   120*

Summary and Current Debate   123

7. Imitation, Self-Recognition, and the Theory of Mind   126

Is There Evidence for Imitation in Animals?   127

The Attribution of Mental States in Animals   132

Self-Knowledge and Self-Recognition   141

*Relationships between Mirror Recognition, Social Attribution,
Imitation, and Teaching   147*

Summary and Current Debate   152

8. An Agenda for Comparative Cognitive Studies   155

Cognitive Ethology: Mental Representations or Mental Experiences?   155

The Generalist versus Ecological Approach to Animal Cognition   163

Conclusions   170

References   175

Index   203

Preface

My goal in this book is to offer the reader a broad perspective on the scientific study of cognition in animals. The question of how animals "think," which has its roots in both biology and psychology, is a major concern of the field of comparative psychology. I focus on this question as it is taken up by those scientists, psychologists *and* biologists, who look for answers mainly through experimentation—that is, who utilize the methods and the techniques of experimental psychology and ethology. My emphasis on experimentation does not imply that only laboratory work is considered here. Indeed, a number of investigators have ingeniously adapted experimental techniques to the field. The domain of social cognition furnishes several examples of experimental investigations of wild animals in their natural environments.

A productive and influential group of biologists is interested in the behavior of organisms in their natural environments. This speciality is known as *ethology* and has developed through the seminal work of scientists like Nikolaas Tinbergen and Konrad Lorenz (see Manning and Dawkins, 1992, for an introduction to ethological studies of animal behavior). Ethologists have studied the nature of the information processed by animals (kinds of external stimuli) and the various forms of responses to these stimuli. Their contributions have been particularly strong in the domain of communication and social organization in various species, ranging from insects to nonhuman primates. For example, a well-known illustration of communication between animals, the honey-bee dance, was described by Von Frisch (1967). Von Frisch has explained how scout bees, after they have discovered a food source, transmit information about distance, direction, and the nature of the food to foragers.

The subfield of communication and social cognition in general is very active in ethology today (see Chapter 5). Behaviors that may be similar to human deception have been described and analyzed in studies of birds (Ristau, 1991a) and other nonhuman species, mostly primates (Mitchell and Thompson, 1986). Other studies have explored the intricacies of primate social organization, such as the formation of coalitions in natural or seminatural environments (Harcourt and de Waal, 1992) or reconciliation among members of a social group (de Waal, 1989).

The stance taken here is somewhat different from, although highly complementary to, the ethological approach. Even though some early ethologists—Tinbergen, for example—made use of controlled experiments, it can be said that ethologists have for a long time relied upon descriptions of the behavior of wild animals. In contrast, the field known as *comparative psychology* typically uses controlled environments (usually under laboratory conditions) to investigate "intelligent" behavior in animals. This discipline, like ethology, has its roots in Charles Darwin's conception (e.g., Darwin, 1871) of a possible continuity in mental capacity among animal species (including man). The program of comparative psychology, which has been progressively elaborated after Darwin, is thus an attempt to grasp the origins and the development of the *human* mind by studying the behavior of *nonhuman* animals. Broadly speaking, the "mind" in this context can be defined as "the set of cognitive structures, processes, skills, and representations that intervene between experience and behavior" (Roitblat, 1987, pp. 1–2).

The first chapter describes the emergence of this field at the end of the nineteenth century, even though the questions related to animal intelligence are as old, in Western culture, as philosophical interrogation itself. For example, ancient Greek philosophers like Aristotle or Plato showed an interest in the intelligence or the soul of animals, whereas René Descartes, in the seventeenth century, denied the possibility of an animal mind.

The historical overview is followed by a discussion of the state of modern comparative psychology and its reliance on concepts, such as representation, proposed by psychologists studying human cognition. In Chapter 1 I also comment on the lack of direct access to cognitive processes that is inherent to this field of inquiry. Cognitive processes are not directly visible; they are accessible to us only by interpreting data gathered in an appropriate context of inquiry. The criteria used to assess cognitive processes must therefore be made explicit.

Chapter 2 deals with the methods and principal topics that define the field of the experimental analysis of animal cognition. Among those topics several examples, concerning categorization and imagery, are presented in this chapter for different species, from monkeys to pigeons and rats.

Chapter 3 is dedicated to Jean Piaget's theories about the development of human intelligence. The application of Piagetian concepts (object permanence, inferential reasoning, numerical competence) to several animal species is described and critically evaluated. Reference to the Piagetian framework supplies the field of comparative psychology with a unique tool for relating activities performed by human infants and children to the performance of animals. In this respect, the Piagetian schema can be used to organize the data and the concepts of the comparative approach.

Chapter 4 addresses the question of spatial and temporal representations and that of tool use. In effect, the cognitive competencies underlying spatial abilities represent some of the most impressive and well-documented cases of cognition in the animal kingdom. Application of these abilities is evident not only in the way different species orient and move their bodies in space but also in the manner they use objects such as tools. In addition, tool use by animals is examined here for what it tells us about its organization and function.

Chapter 5 is concerned with the problem of social cognition. It starts with a consideration of the methodological constraints posed by the study of social cognition in the laboratory (for instance, finding the best conditions for recording interactions between two or more individuals). This chapter reports several studies, mostly investigations of nonhuman primates, dealing with mother-infant relationships, dominance, and social affiliation. The possibility that animals have built social "concepts" through interactions with conspecifics is reviewed, as is the relation of social cognition to nonsocial cognition (such as the knowledge of inanimate objects).

Chapter 6 is devoted to the examination of communicative abilities of animals and a comparison of these skills and human language. Several communicative systems are explored, ranging from the honey bee's waggle dance to the signaling behaviors (for example, the alarm calls) of birds and nonhuman primates. Characteristics of the animal signals are related to features of linguistic abilities. Special emphasis is given to "language-trained" marine mammals (sea lions and dolphins) and apes (particularly the chimpanzee). The different media (gestural communication, tokens) used by researchers in their attempts to teach animals some of the rudiments of language are examined. After a proper description of these studies and of their main findings, the characteristics of the behaviors produced by the trained animals are compared with verbal signs and particularly with the acquisition of verbal signs by humans.

Specific functions that are found in human communication but (probably) not in animals are then considered. Examples are the distinction between imperative and declarative functions and questions of predication, grammar, and "radical arbitrariness." This section explores the prerequisites of verbal signals in the context of prelinguistic communication, object manipulations, and social and object-oriented exchanges between infants and competent adults.

Chapter 7 deals with other important topics in the study of the acquisition and transfer of knowledge between two or more individuals—namely, imitation, self-recognition, and the "theory of mind." The evidence for imitation in animals makes explicit the association between this ability and underlying cognitive processes. The questions of tradition and culture are approached through some well-described cases (e.g., potato washing in Japanese macaques).

Two different views can be adopted in studies of animal cognition. The first is called "information-processing psychology." It uses the cognitive concept of representation to study how information is encoded by organisms and to analyze the transformations that are applied to coded representations. Some authors have described this approach as the study of the "computing mind" (Prato Previde, Colombetti, Poli, and Spada, 1992). This framework underlies the analysis of cognition presented from Chapter 1 to Chapter 6. The second approach, encompassing most of the findings reported in Chapter 7, has been coined the study of the "semantic mind," because it regards representations as mental states—that is, as sets of beliefs, desires, or intentions about objects, facts, and events in the environment.

This rather recent interest in the attribution of mental states in different animal species is detailed in Chapter 7, which includes a discussion of instances (in birds and nonhuman primates) whereby information is manipulated during communicatory exchanges (for example, the concealment of information and deceptive behaviors). Several recent studies are reported documenting that apes (namely chimpanzees), but not monkeys, are capable of some form of attribution of knowledge to others (chimpanzees or humans). This attributive skill is discussed in the context of animals' ability to recognize their reflections in a mirror, to guide their activities through indirect media, and to control their subjective states. This chapter then proceeds to bring into one single framework the relationships among mirror-recognition, social attribution, and imitation. The consequence of possessing attributive capacities (or a theory of mind) for teaching abilities is explored by documenting the scarce evidence for pedagogy among animals.

Chapter 8 offers an agenda for comparative cognition. It reviews and synthesizes the findings brought up by the study of cognitive processes in animals. Here I offer a description of very general levels of processing as well as types of representations. Different levels of representational systems can already be described and compared with empirical findings. In turn, these levels can serve as indices to differentiate the types of cognitive processing used by organisms to respond to their physical and social environments. Among the hypotheses advanced in this section is a distinction between a representation and a rule of action.

Scientists involved in the study of animal cognition can be categorized broadly as belonging to one of two groups, each having a specific program: the ecological program or the generalist program. The distinction is in the approach one takes: the field versus the laboratory. The merits and limitations of each program are delineated in Chapter 8, and the possibility of reconciling the two and taking an integrative approach to cognitive processes in animals is examined.

Chapter 8 also introduces approaches to the study of the animal mind other than that derived from the experimental analysis of comparative cognition. These approaches are sometimes complementary to that of comparative cognition, but distinct conceptions are sometimes more or less radical alternatives to the mainstream approach. The "cognitive ethology" developed by Griffin (e.g., Griffin, 1992) is an example of the latter. This theory is briefly presented and its concepts and purposes are examined in light of the experimental approach (that is, the generalist program).

Chapter 8 ends with a portrait of animals based on what we have learned from studies of their cognitive achievements, that is, as problem solvers and decision makers. If this view of animals as active agents able to adapt to their environment holds true, then we must reconsider, in the light of contemporary research, the debate about the continuity of mental functions between animals and humans.

Throughout the text I have referred to animals by their common names rather than by the scientific names of species. This choice reflects the fact that scientific names are not systematically provided in the literature, though the interested will find many species names in the references I have cited.

Most chapters conclude with a summary and a brief discussion of current debates in the field. There is much controversy over crucial questions such as the impossibility of proving the existence of consciousness and elements of discontinuity between humans and nonhumans. Suggestions for further reading concerning specific domains that have been touched upon only lightly in the chapter are likewise provided.

Portions of this volume are based on my book, *L'intelligence de l'animal,* which was published in French by Editions du Seuil (Paris, 1992).

I am indebted to James R. Anderson, Roger K. Thompson, Matthew J. Jorgensen, and Frans B. de Waal for their helpful comments and suggestions for improving the manuscript.

Animal Cognition

1

Origins and Development of the
Study of Animal Cognition

Experimental approaches in contemporary studies of animal cognition derive from two main sources. The first source is the Darwinian tradition in biology (Darwin, 1871), according to which humans share a number of cognitive abilities with other living species. The second source is human cognitive psychology, which has provided useful concepts and models (such as the concept of representation or the theories of information processing: Wasserman, 1981; Yoerg and Kamil, 1991) as well as experimental methods (such as mental chronometry: Posner, 1978). In this chapter, I discuss these roots of the discipline known as "comparative psychology" and how they have influenced recent approaches to the study of the animal mind.

The Darwinian Heritage and Nineteenth-Century Psychology

Darwin's conceptions about adaptation and natural selection have played a seminal role in shaping the hypothesis of continuity between animal species and the human species. This idea of continuity was so obvious to Darwin that he postulated only quantitative differences between humans and the other species: "The difference in mind between man and the higher animals, great as it is, certainly is one of degree and not of kind" (Darwin, 1871, p. 128).

According to this frame of reference, the human species was understood as the product of ancient forms. The mechanisms responsible for evolution shaped not only a species' anatomical characteristics but also its psychological traits, called "mental faculties" by Darwin. These "faculties" referred to curiosity, imitation, attention, memory, and reasoning. Even though Darwin accepted the existence of "beliefs" in animal species equivalent to those found in humans, he remained cautious about attributing language and consciousness to any higher-order nonhuman species. Nevertheless, he considered that some elements of language were already present in animal communication (bird songs and primate vocalizations) and that those elements, when combined with novel mental abilities, would inevitably lead to human language.

In the seventeenth century the French philosopher René Descartes established the idea of dualism, or discontinuity, between the human (the mind) and the animal (the machine). Darwin's emphasis on continuity among all species brought some "humanity" back to the animals. More important, his ideas have made the study of animal behavior an essential tool for understanding human behavior. In effect, if one accepts that the human species is the product of some ancestral nonhuman forms, then the study of animals' mental functions becomes indispensable for understanding the biological precursors of the human mind.

The mental life of animals was considered by the first comparative psychologists soon after Darwin's work appeared. For example, George Romanes published *Animal Intelligence* just a decade after Darwin's *Descent of Man.* Toward the end of the nineteenth century, the method of introspection was the main technique of investigation in human psychology. Given that this method is impossible to use with animals (because of their inability to explain to us their inner states), Romanes developed another technique, "subjective inference" (Wasserman, 1981). This method consists of projecting conscious experience onto other organisms. Because animals cannot "tell" us their internal states, according to Romanes, we need to make the inference that their activities are analogous to human activities and that the mental state corresponding to an activity performed by a human is also present in an animal performing the same activity.

A major contribution of this approach was to place animal mental functions within the context of learning abilities. To Romanes, an organism can make novel adjustments or change old adaptations as a result of its experience. This interest in learning capacities was further developed by other comparative psychologists, including Morgan (1894), Loeb (1900), Washburn (1908), Jennings (1906), Thorndike (1911), and Yerkes

(1911). General psychology was also influenced by the Darwinian perspective. For example, Chicago-based psychologists concerned with the function of mental phenomena (such as James Rowland Angell, Harvey Carr, and John Dewey) viewed animal thinking as a set of adaptative processes. This functional orientation promoted cross-species comparisons as a means of fully understanding the functional aspects of all behaviors (Thompson and Demarest, 1992).

Positions such as those defended by Romanes are a prime target for those psychologists who demand purely objective descriptions in animal psychology. Indeed, these positions were strongly criticized by Watson (1919) and by the behavioristic movement.

The Behaviorist Break

The approach of the first comparative psychologists—assuming correspondence between human and nonhuman mental states—risked criticism because of its anthropocentric stance. The first to react to this drift was C. L. Morgan, who warned against the danger of interpreting an action by reference to a higher mental faculty when it could be explained by lower-order explanatory schemas. This recommendation is known as "Morgan's canon" or "principle of parsimony." Morgan stated that "in no case may we interpret an action as the outcome of the exercise of a higher psychical faculty, if it can be interpreted as the outcome of the exercise of one which stands lower in the psychological scale" (1894, p. 53). Consider the case of a butterfly's orientation toward a source of light. The parsimony principle would favor an explanation in terms of the action of a preprogrammed mechanism (called "tropism") that attracts the insect toward lighted spots, rather than an interpretation invoking, for example, the butterfly's curiosity.

Morgan's work began a new direction in animal behavioral studies, one that aimed at an objective description of the phenomena the biologist or the psychologist tried to grasp. This current was strongly developed by several laboratory studies carried out by Edward Thorndike (1898, 1911). Thorndike pioneered the use of "puzzle boxes" and mazes to study species such as rabbits, cats, and rats.

Animal psychology was also deeply influenced by the work of J. B. Watson (1919). Watson firmly condemned the studies of early comparative psychologists and their attempts to describe the life of animals in its totality, including animal consciousness. For Watson, who proposed a strictly descriptive approach, the study of animal behavior should be limited to relations between environmental stimulations and reactions of organisms to those stimulations. In fact, the project pursued by Watson specifically excluded mentalism from psychology. His goal was for psychology to attain a status comparable to physics—that is, a scientific discipline that took into account observable phenomena only. In other words, any speculation about the functioning of the "black box"—namely, the processes that might mediate between sensory inputs (the stimuli) and behavioral outputs (the responses)—is forbidden.

This sort of reductionism has had several consequences for experimental animal psychology. Most important is the fact that for decades any reference to internal activities of the organisms was discouraged.

The Emergence of the Cognitive Approach

The Delayed Response

Despite the behaviorist stance, several attempts were made to study the behavior of animals in a "mentalistic" way. For instance, Hunter (1912) designed an apparatus for the study of *delayed reactions* in rats and dogs. In one of Hunter's experiments, the animal is placed in a starting box, which is separated by a window from the rest of the cage. Three different doors arranged in a half-circle can be reached from the starting box. The entrance to each door is indicated by a light bulb. The activation of the light bulb signals the presence of food behind the door. The light corresponding to the baited compartment is briefly turned on. After it is turned off and following a variable delay (up to ten seconds), the animal has free access to the apparatus and gets the reward if it goes directly to the correct door. Under these experimental conditions, the animal must produce its response in the absence of the stimulus. Hunter interpreted the delayed reaction to an absent signal to mean that the animal responds to a substitute for the stimulus, or, in other terms, to its representation or memory.

The experimental framework proposed by Hunter was used and extended by Tinklepaugh (1928, 1932) in studies with monkeys and chimpanzees. These experiments are interesting on several grounds. First, Tinklepaugh correctly foresaw that the kind of stimulus used in Hunter's

experiments as indicators for food might not be the most appropriate cues
for the vertebrates studied. For instance, dogs need to perform hundreds
of trials before they can reliably utilize visual cues. Starting from the
observation that dogs and other animals will spontaneously search in
places where they have previously seen food, Tinklepaugh (1932) suggested
using a relatively simple method for the investigation of delayed responses.

An experimenter places two identical cups in front of the subject, one
of which is baited, and then a screen is interposed between the animal and
the two containers. After a variable delay, the screen is raised and the
subject can freely move toward the containers. This apparatus and procedure require that the subject form and keep in memory a "trace" of the
baited cup. This way of presenting spatial problems can of course be made
more complex by adding new containers. Tinklepaugh carried out experiments with macaques and chimpanzees with 16 different pairs of objects,
a piece of food being hidden under one cup of each pair. Even though the
species tested differ in their ability to retrieve the correct cup (on average,
66 percent correct responses for macaques and 90 percent correct responses for chimpanzees), both succeeded in performing the task. For
Tinklepaugh, this suggests that the animals make their choices on the basis
of a representation of the position of each container and not on sensory
guidance. The role of sensory cues was further controlled by performing
a slight rotation (to the right or to the left) of the entire configuration of
cups, which did not affect the subjects' performance. It may finally be
noted that these delayed-response experiments, which require the search
for a hidden object, predate by a few years the studies of "object permanence" by Piaget (1954) in human infants (see Chapter 3).

The delayed-response method was further developed in the United
States by Maier and Schneirla (1935), who proposed a systematic treatment of animal behavior and psychological problems from bacteria to
mammals.

Learning and Problem Solving
A chief defender of the cognitive approach to animal thinking was E. C.
Tolman (1932). Tolman believed that the organism plays an active role in
its adaptation and learning and that behaviors are goal-oriented. Some
important concepts were introduced by Tolman in the fields of experimental and comparative psychology. Two of those, *latent learning* and
cognitive maps, have had a lasting influence on our conceptions of learning

formation and the organization of spatial orientation. In a famous paper of 1948, Tolman defined cognitive maps as follows: "We believe that in the course of learning, something like a field map of the environment gets established in the rat's brain and the incoming impulses are usually worked over and elaborated into a tentative, cognitive-like map of the environment" (p. 192). A cognitive map is formed of two related pieces of information: first, locations and their connecting paths and, second and simultaneously, objects or events associated with the given space.

The phenomenon of latent learning illustrates this twofold characteristic of cognitive maps. In latent-learning experiments, rats are able to learn where food and water are located in a maze, even if the rats are neither hungry nor thirsty (Spence and Lippitt, 1940). The rats are free to explore a Y maze: one end of the maze contains a cup with food, the other end contains a cup with water. After the exploratory phase, rats are divided in two subgroups: one group is food-deprived, the other water-deprived. Each rat is then left at the entrance of the maze. The results are clear-cut: thirsty animals orient directly toward the water cup while hungry rats move toward the food cup. This finding shows that during the exploratory phase, rats have learned some of the features of the maze independently of the satisfaction of immediate needs (as implied in the behavioristic model of animal learning). These features constitute elements of the rats' "cognitive maps," which include both locations in space and associated events or objects (Tolman, 1948; Vauclair, 1987).

Tolman's cognitive approach faced vigorous competition, in particular from stimulus-response theorists such as C. L. Hull (Amsel and Rashotte, 1984). For example, Hull (1934) proposed alternative explanations for learning behaviors, such as "habit family hierarchies" and "fractional antedating goal responses." Tolman's cognitive maps became more popular than Hull's explanations, however, probably because, as noted by Walker (1983), "Hull's system seems to require more cognitive effort by both the human theorist and the rats" (p. 77).

Before closing this historical section on the precursors of the analysis of cognitive processes in animals (see Table 1.1 for a summary, and see Boakes, 1984, and Richards, 1987, for a review), I must mention the pioneering work of Wolfgang Köhler (1925). This researcher studied problem solving in chimpanzees, such as their use of tools to reach otherwise inaccessible objects (see Chapter 3). Köhler is also at the origin of the concept of "insight" used to explain the sudden recognition of a solution

Table 1.1. Precursors to contemporary studies of animal cognition.

Field of study	Representative figures
Theory of evolution	Darwin (circa 1871)
Comparative psychology	Romanes, Morgan
Behaviorism	Watson, Skinner
Cognitive animal psychology	Hunter, Tinklepaugh, Tolman
Human cognitive psychology	Neisser, Simon, Piaget
Experimental animal psychology	Premack, Terrace
Cognitive ethology	Griffin

to some cognitive problems. Köhler and the school of Gestalt psychology, which has its origins in Germany, strongly influenced cognitively oriented investigators in the United States (like Tolman) and in Europe (Guillaume, 1940).

The Modern Concepts of Representation and Memory

A major impetus for the development of the cognitive approach in the field of animal research has come from the cognitive sciences and, within them, from the psychology of human cognition. Cognitive sciences have developed with the rise of information theories concerning the capacities of human beings and computers to process information (Johnson-Laird, 1988).

A central concept in cognitive theories is that of representation (Neisser, 1967). Briefly stated, this concept refers to the fact that information gathered by an organism is organized in the form of "internal representations." The formation of representations involves the intervention of some kind of computations or combinatorial procedures acting upon the results of sensory and perceptual processing (Gallistel, 1990). This information-processing approach to cognition implies, in effect, that mental events are to some degree analogous to informational events. "Information in this system is embodied in states called representations; the operations that are performed on that information are embodied in changes of state called processes. The task of the cognitive psychologist from an information processing perspective is to determine the nature of the processes

Table 1.2. The behaviorist and cognitive conceptions of learning.

Behaviorist model:	Stimulus (S) → Response (R)
Cognitive model (based on information theory):	Input → Operation → Output

which transform, encode, represent, and use information from the external (or internal) world to produce behavior" (Yoerg and Kamil, 1991, p. 279).

The traditional view of animal learning is that learning is formed from a relation between a stimulus (S) and a response (R). Within the information-theory perspective, this S-R relation is replaced by a different string of components (see Table 1.2): an input of information is substituted for "S" and an output of the system is substituted for "R." The inputs of information are contained in states called *representations,* whereas *processes* are responsible for the changes in states that lead to the outputs (overt behaviors) of the representational system. The S-R model stresses the role of the stimulus (the information coming from the environment) and of the response (the action produced by the organism). In contrast, the cognitive model stresses the interplay between the stimulus and the response and leads to reorganizations and generalizations of the input and output systems, which are largely independent of the specific stimulus and response.

The Study of Representation in Animals

If one accepts the approach described above to cognition in general and to animal cognition in particular, one has to consider animals not as reflex automatons but as systems that process information in order to adapt to their environment. Various cognitive abilities intervene in this adaptation, such as perceptions, learning, memory, and problem solving.

At first glance, the behaviorist approach would appear very remote from cognitive psychology, but it has in fact largely contributed to its development. As a matter of fact, most leading figures in the field of cognitive animal psychology come from the behaviorist tradition. These scholars have brought a methodological rigor to their investigations of animal cognition; moreover, they have adapted the experimental techniques developed in the study of animal behavior to the cognitive framework: see, for example, the book *Cognitive Processes in Animal Behavior* (Hulse,

Fowler and Honig, 1978), which represents a significant effort in the rebirth of animal cognition studies.

If cognition is essentially tied to the processing of information into representations, a major task for the animal psychologist is to design appropriate experiments for showing if and how an organism is able to activate and use information that is not available in its present environment. A representational activity thus implies an ability to form a trace of a stimulus or, in other terms, to encode a stimulus in memory and then to use this trace to react appropriately to current environmental conditions.

Functional differences between *memory* and *representation* are hard to make, and in fact both terms are often used interchangeably. For some authors, memory is defined as "an internal representation of some event an organism has experienced in the past" (Gordon, 1983, p. 399). Using representations or memories to explain how animals adapt to their environment is not all that mysterious when one realizes that the ability to construct and use a representation confers obvious adaptive advantages to the organism that possesses it. For example, an animal without some sort of spatial representation of its familiar surroundings would face terrible difficulties in locating its home area and other sites where food or mates are to be found. Similarly, an individual could hardly survive within a social group without some sort of representation of the other individuals that compose the group.

The Problem of Indirect Access to Representations

Contrary to the pure behaviorist tradition, according to which stimuli and responses must be directly observed and measured, the study of representational concepts calls for a method of indirect evaluation or, in other words, an *inferential process*. These inferences about the structure and function of representations are based upon an analysis of the effects representations have on behavior. Thus, because cognition involves learning and thinking processes that are not directly observable, it is necessary to design experiments that can elicit a response from the organism to demonstrate the use of a previously perceived and stored representation. In this respect, a number of criteria are used to assess the cognitive nature of the produced response.

The use of the concept of representation or more generally of representational systems in animal cognition calls for a relation of mapping or isomorphism between some or all features of the world and (some or all) features of the representation (Roitblat, 1982, 1987; Gallistel, 1990).

Another aspect of representations that must be stressed, especially in the context of the study of animal cognition, is that the rationale for studying representations and cognitive processes in animals requires no reference to animal consciousness (Terrace, 1984).

Problems Posed by the Study of Cognition in Animals

The notion of "cognition" has replaced and extended the more general concept of "intelligence" that had been championed by Darwin's followers (e.g., Romanes, 1882). In fact, it appears that the terms have been systematically confused. McFarland (1989) has suggested as a remedy that we make a distinction "between cognition, a possible means to an end, and intelligence, an assessment of performance judged by some functional criteria" (p. 130). The example of pigeons' navigational abilities (Wiltschko and Wiltschko, 1987) serves to illustrate this distinction. These birds are able to find their home by using different navigational systems, such as a sun compass or sensitivity to magnetic fields. These systems are very sophisticated and they would probably look highly "intelligent" if performed by a robot. The pigeon's navigational capacities are controlled mostly by hard-wired systems, but these systems are not generalizable to other domains. According to Rozin (1976), intelligent behaviors are conceived of as specific adaptations to specific problems. The pigeon has various and complementary orientation systems to return to its loft, but these abilities do not seem to generalize to other domains in relation to its adaptation to the environment (for example, in the search for food or partners; but see Chapter 8). The concept of intelligence must thus be restricted to performance evaluation with respect to a given criterion. By contrast, the concept of cognition is reserved for the manifestation of learning and information processing within a given individual. Appropriate criteria allowing the identification of cognition must be used to distinguish these processes from behavioral organizations, complex or not, that result from hard-wired rules in the nervous system.

In general terms, cognition is seen to allow an individual to adapt to unpredictable changing conditions in its environment. Thus, behaviors that would aid adaptation would reflect several characteristics, such as *flexibility, novelty,* and *generalization*. Flexibility of a behavior designates the possibility of constructing an adapted response to unusual external conditions. The response must also be novel in the sense that it does not express the existence of a prewired program. Finally, the novel behavior, established to solve a novel problem, must be susceptible to generalization to situations that differ partially or totally from those in which they were initially acquired. The following chapter will provide examples of cognitive behaviors fulfilling one or more of these characteristics.

2

Laboratory Methods for Assessing Representation in Animals

The possibility that representational processes are at work in animal cognition may be assessed in a variety of problem-solving situations, such as those requiring discriminative learning and perceptual categorization. Problem-solving behaviors express the control the animal has over the dimensions (physical attributes) of the processed information, as well as its mastery of the abstract relations applied to these dimensions.

An important objective of the study of representation in animals is to determine the level of abstraction at which the organism processes the stimuli. The question is whether a stimulus is exclusively processed by taking into account its physical attributes, or whether the organism can process relations between stimuli independently of their concrete dimensions (size and color, for example). Several studies illustrate the level of abstraction involved in the coding of information by animals. In this chapter I discuss these studies and the explanations of problem solving that have been based on their results—namely, the strategies expressed in learning sets, the construction of the "concepts" of identity and oddity, the elaboration of perceptual categories, serial learning, and imagery.

Learning Sets

Macaque monkeys that have been trained on hundreds of trials to make visual discriminations between two objects are able to make novel discriminations after one trial only. In these experiments (Harlow, 1949; for a review, see Fobes and King, 1982), the macaque is first trained to discriminate between two visual stimuli by being systematically rewarded with food if it chooses the correct one. After a number of trials with the same pair of stimuli, a new pair is introduced. The number of trials for this novel pair is identical to the number of trials required to train the monkey to discriminate between the first pair. Training continues in this way until the monkey has learned hundreds of different pairs. This procedure is selected to prevent the monkey from forming too strong an association between specific stimuli and reinforcement. Correct performance of the macaques increases with the introduction of more novel pairs. For example, the percentage of correct responses for the first 8 discriminations is about 52 percent but reaches 100 percent after 300 different discriminations have been made.

According to Harlow, the monkeys have learned a rule that they are able to generalize to the novel pairs of stimuli; in short, they have learned how to learn. In other words, learning does not consist of the formation of simple connections between the stimuli and the response(s); instead, it involves the construction of a method, of a learning strategy, called a *learning set*. It is assumed that the animal has built a representation of the chosen stimulus during a given trial and of the consequences (reinforcement or not) of its choice. This representation is sufficiently abstract to allow the subject to apply it to any pair of novel stimuli. The use of such a rule explains, in Harlow's view, the fact that learning is not limited to making a simple connection between a particular stimulus and response and that it can occur in one single trial. The ability to learn a behavior on the first trial, called "insight" by Köhler (1925), could in fact be a consequence of prior learning situations.

Mastery of Relations between Stimuli

An important dimension in learning is the capacity to use identity ("same" concept) and nonidentity relations ("different" concept) between objects and events. The capacity to detect and use such relations is demonstrated through a procedure known as matching-to-sample (MTS). The MTS procedure involves the presentation of a sample stimulus and of two (or more) test stimuli, one of which is identical to the sample. The sample and test stimuli may be presented simultaneously or with a time interval. Two experimental procedures may be followed: either the choice of the stimulus that is identical to the model is reinforced (this is the MTS task), or the choice of the different stimulus is reinforced (the non-MTS, or NMTS, task, also known as the "oddity-from-sample" task). The general methodology consists of training the animal with a limited number of stimuli (e.g., blue and red objects), and once performance reaches a consistently high level of correct responses, the subject is tested with novel stimuli. Pigeons are able to respond to visual stimuli on the basis of a relation of sameness, as in color or form (Zentall and Hogan, 1976), but thousands of trials are needed for learning to take place (Santiago and Wright, 1984). Mammals seem to be more adept than pigeons at identity and oddity learning (for an example with auditory stimuli in the dolphin, see Herman and Gordon, 1974; for an example with visual stimuli in the monkey, see D'Amato, Salmon, and Colombo, 1985).

Each generalized MTS trial entails but a single identity-equivalence judgment, but the matching concept can be more complex than just expressing an identity. For example, MTS tasks might be used to test a subject's ability to judge whether "relations between relations" (Premack, 1983) are equivalent (Thompson, 1995). One might test whether a pair of balls is perceived or judged to have the same relation (identity) as a pair of shoes, for example, or whether a paired lock and cup have the same relation (nonidentity) as a paired pencil and canister (Oden, Thompson, and Premack, 1990). The data from the literature indicate that only humans beyond infancy and adult chimpanzees with a history of language training (see Chapter 6) can explicitly judge relations (same or different) about relations (identity and nonidentity: Thompson, 1995). Moreover, adult rhesus macaques, unlike the child and the chimpanzee, do not spontaneously perceive abstract relational similarities and differences despite their sensitivity to physical features when tested with the same task (see Thompson, 1995).

The level of competence attained by these species in mastering the relations of resemblance and difference implies the elaboration of representations related not only to the objects and their physical parameters but also to abstract relations between them. These abilities are further exemplified in the discrimination of categories.

Category Formation

Just as the construction of relations of similarity or difference between the physical characteristics of stimuli implies reference to an abstract representation of the identity relation, the capacity for categorization implies reference to an abstract representation of the class itself; this class must be discriminated as an entity that is different from another entity (i.e., another class).

Natural category learning has been studied in the pigeon (e.g., Herrnstein, Loveland, and Cable, 1976; Herrnstein, 1979). Subjects were presented with a set of 80 slides (one at a time), 40 containing trees and 40 without trees. Pecking to a slide depicting a scene with a tree (defined as the positive stimulus) is reinforced with a reward of food, and pecking at a slide without a tree is never rewarded. The slides could contain objects other than trees, as positive stimuli, or the trees could be represented as silhouettes or be partially obstructed. After a relatively large number of trials, most pigeons were able to reliably discriminate tree versus non-tree pictures. Moreover, the discrimination generalized to new displays of trees with little or no decrease in competence from the subject's performance on training slides. The authors concluded that the pigeons used an abstract representation—in other words, a "concept"—of a class of natural objects (trees) to make discriminations.

The ability to form categories is apparently not limited to the identification of objects, like trees, that possess an obviously adaptive significance for a bird. Pigeons are also able to distinguish underwater scenes with fish from the same scenes without fish (Herrnstein and de Villiers, 1980), as well as other natural (humans or pigeons) and arbitrary categories (letter *A* versus number *2*: Morgan, Fitch, Holman, and Lea, 1976).

More recent studies have confirmed the ability of pigeons to master categories. For example, pigeons have been trained to classify objects belonging to four different categories (cats, flowers, cars, or chairs), by pecking one of four keys surrounding a viewing screen onto which slides are displayed (Bhatt, Wasserman, Reynolds, and Knauss, 1988). After 30 days of training, the pigeons reached a criterion of 80 percent correct responses. After passing this acquisition stage, the subject was given a generalization test consisting of completely novel slides (10 for each category). The pigeons solved the task with an average of 64 percent correct responses, suggesting a limited capacity for generalization of their categorization ability.

In spite of these results, it would certainly be unwise to conclude that the pigeon has abstracted a singular property or a *prototype* of the object (for more on "prototype theory," see, e.g., Rosch, 1978; but also see Wasserman, 1993), as a human subject would do when confronted with a similar discriminatory task. In effect, before postulating that the pigeon can master the concept of an object (natural or abstract), we must first consider a simpler hypothesis: that performance is controlled by a combination of perceptual features ("feature theory") and that the category is polymorphous. Results from the study by Morgan, Fitch, Holman, and Lea (1976) support such an interpretation. After being trained on the discrimination between A and 2, pigeons were given a transfer test in which all the other letters of the alphabet were presented. Letters having their apex at the top and two bottom projections were responded to as the A stimulus was (the positive stimulus to which pecks were directed), while letters with curves or loops were treated as a 2 was (the negative stimulus). For further discussion on categorization of visual stimuli by animals and on human categorization processes, see Herrnstein (1990).

Categorization processes in pigeons and in other animals (Schrier, Angarella, and Povar, 1984; Roberts and Mazmanian, 1988; Gardner and Gardner, 1984; Vauclair and Fagot, 1996) are far from being completely understood. Data gathered so far suggest, however, that the animals use relatively abstract classificatory systems based on object representations that are more complex than the simple physical parameters of the objects. For a critical review of conceptual knowledge in animal cognition studies, see Thompson (1995).

In human cognition, most representations are linguistically based, but even for us there are several domains for which nonverbal thinking is at work. The rest of this chapter focuses on forms of nonverbal thinking, such as list or serial learning, memory, and imagery in animal species. Data from similar experiments using birds, monkeys, and humans will allow comparisons between human and animal performance as well as comparisons between different nonhuman animals.

Serial Learning as Evidence of Nonverbal Thought

Representation in List Learning

List-learning studies require subjects to learn a list of simultaneously chained items (Straub and Terrace, 1981). In a typical experiment, a pigeon is presented with an array of four colors (A, B, C, D). In order to get a reward the subject must respond (by a peck) to each item in the sequence (A-B-C-D) regardless of how the colors are positioned on the response keys. The subject gets no differential feedback after each correct peck but, rather, must correctly complete the entire sequence before being rewarded with food. A novel configuration of list items is used from trial to trial to prevent the use of a unique chain of physical responses in sequence production. These procedures ensure that "decisions regarding the next item to be selected must be based on the ability to use some representation of the sequence that defines the subject's current position in the sequence" (Terrace, 1993, p. 163).

Studies have shown that pigeons learn to perform correctly with lists of three and four items and that they perform just as well with configurations in novel arrays. For Straub and Terrace (1981), the pigeon has mastered a routine acquired during learning, namely a representation of ordered colors. This representation of the list could explain why the birds adjust so easily to novel arrays.

This same ability is further demonstrated by the pigeon's accurate performance on two-item subsets that can be derived from the list (such as the subset AB, AC, AD, BC, BD, and CD). Response accuracy is high across all subsets except for subset BC. The specific difficulty encountered by the pigeons with this subset is important to note. The mastery of the relations pertaining to lists will be discussed in the context of inferential reasoning in Chapter 3.

These results reinforce the interpretation that the pigeon builds, during learning, a representation of the list and that it explores this representation to produce a novel list. An analysis of latencies of correct responses to the first and second items of each subset may explain the specific difficulty encountered with the subset BC. This analysis shows that the pigeon is, on average, twice as fast to respond when the subset begins with the item A than with either B or C. From accuracy and latency data, three response rules can be described: (1) respond first to item A; (2) respond last to item D; (3) respond to any other item by default. Thus, rules 1 and 2 cannot apply for subset BC, since this list contains neither A nor D.

The simultaneous-chaining procedure has been applied in studies with capuchin monkeys (D'Amato and Colombo, 1988). First of all, the monkeys acquired a 5-item list faster than pigeons did, but more interestingly their performance on subsets of the original list indicate their representation of the list is different from the pigeons' representation. This difference is revealed by differences in the latencies of the responses to items in the first position of each two-item subset.

The latency to respond increases monotonically in monkeys when the position of the item on the original list moves from the initial to the final position (see Figure 2.1). In pigeons, response latency remains constant. Terrace (1993) concludes that the monkey develops a linear representation of the list: "In order to decide which member of a pair to respond to first, a monkey starts at the beginning of its representation of the list learned and moves through it until it locates one of the items displayed on the subset test. The more items the monkey has to check, the longer the response latency. By contrast, the pigeon appears to rely on its representation of the highly salient first and last items of the list and a default rule to determine the order of the items it pecks on a subset test" (Terrace, 1993, p. 165). These rules could explain the drop in accuracy found when pigeons are presented with subset BC, given that neither of the elements of this subset is the first or the last item of the original list.

In brief, the use of the simultaneous-chaining procedure provides a sensitive tool for illuminating inter-species differences in the nature of the representations intervening in this kind of task.

The Effect of Serial Position

Studies with human subjects have demonstrated a twofold impact of the order of presentation of items in a list to be memorized: initial and final items of the list are generally better recalled than items located in the

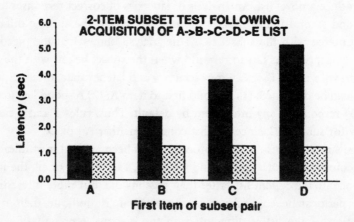

Figure 2.1. Mean latency of responses to the first of two items in a test pair as a function of that item's position on the original list. *Filled bars:* data from monkeys. *Cross-hatched bars:* data from pigeons. (From Terrace, 1993; reprinted with the permission of Cambridge University Press.)

middle of the list. These effects are designated, respectively, as "primacy" and "recency" effects, and they are often obtained with linguistic material, such as lists of syllables or words (McCrary and Hunter, 1953).

Several studies with nonhuman species have found a recency effect associated with the absence of a primacy effect. Examples include spatial learning in rats (Roberts and Smythe, 1979) and memory for lists of sounds in dolphins (Thompson and Herman, 1977). Both recency and primacy effects have been observed with rats (Bolhuis and van Kampen, 1988) and with food-storing birds, like the black-capped chickadee (Crystal and Shettleworth, 1994). These effects may, however, be partially dependent on the time elapsed between the caching phase and the recovery phase. Recency is apparent only when this time interval is 2 to 3 hours; it disappears after a 24-hour delay (Sherry, 1984).

Primacy and recency effects have been investigated in monkeys by use of two distinct experimental procedures: delayed matching-to-sample (DMTS) and serial probe recognition (SPR).

In DMTS experiments, the subjects (four squirrel monkeys and four human subjects) are shown lists containing one, three, or six patterns (geometric forms). The sample stimuli are displayed on the center key of three keys on a board. Both the duration of the sample stimulus presentation and the delay between sample offset and presentation of test stimuli are manipulated. After the center key is turned off, a variable delay is introduced (between 0.5 and 5.0 seconds) and the two side keys are illuminated: one key matches the sample pattern and the other is a nonmatching key. The subjects give their response by pressing one of the side keys. A press to a matching side key ends the trial and provides a reward pellet for the monkeys. In this experimental situation, primacy and recency effects are found with both monkeys and humans (Roberts and Kraemer, 1981). Neither variation in presentation time nor increase in the delay alter the position effects observed. In other words, the middle items of the series are less accurately matched than are the first (primacy effect) or the last (recency effect) items.

Further evidence of differences due to serial position comes from a study using the SPR procedure. This procedure involves the successive presentation of a list of items and, after a delay, the presentation of a test stimulus (the probe), which belongs to the list or not. The apparatus designed for a macaque (Sands and Wright, 1980) comprises two separate screens for the projection of slides (pictures of fruits, flowers, people, animals, etc.). Each trial is initiated when the subject presses a three-position lever. Then, items of the list appear in sequence on the top screen. One second after the last list item is shown, a probe appears in the bottom screen and remains in view until the subject makes a response. A movement of the lever to the left indicates a "same" response and a movement to the right a "different" response. The monkey has been trained to memorize relatively long series of items (up to 20). It performs with high accuracy in this SPR task using color slides as stimuli. Moreover, the curves plotting the results demonstrate both a primacy effect and a recency effect that are very similar to results from a human subject tested under the same conditions as the monkey (Sands and Wright, 1980; for additional evidence of these memory effects in primates, see Castro and Larsen, 1992).

Some scholars—such as Gaffan (1992)—have challenged the experimental evidence for the primacy effect on animals' memory on methodological grounds. For example, it seems that the retention interval between list presentation and memory test is a critical variable: when the retention interval increases, recency effects diminish and primacy effects appear (Wright, Santiago, Sands, and Kendrick, 1985). Controversies surrounding this question have been revived in recent papers (Gaffan, 1994; Kesner, Chiba, and Jackson-Smith, 1994; Wright, 1994).

Because of the similarity in memory mechanisms between monkeys and humans, all the experiments described here underline the interest in using nonhuman primates to study memory processes. One potential contribution of a primate model of human memory could be to clarify the degree to which the functional properties of memory for lists are independent of the use of the linguistic code.

Mental Images in Animals

There is a great deal of evidence suggesting that humans are capable of mentally rotating perceived or imagined visual forms (e.g., Shepard and Metzler, 1971; Corballis, 1988). This ability implies a comparison between a previously presented visual sample stimulus and the display of the same stimulus depicted in different orientations. A typical result of experiments using mental rotation tasks is that decision time increases linearly with the angular disparity of the patterns. It has been suggested that the decision time increases with increasing disparity because subjects "mentally rotate" one visual pattern into congruence with the other (Shepard and Metzler, 1971; see also Shepard, 1982; Kosslyn, 1980).

The above example suggests that humans utilize mental images of objects without any direct and obvious reference to language. Consequently, imagery could be a mental activity that is particularly well adapted to inter-species studies. Surprisingly, very few data are available (Rilling and Neiworth, 1987) to enrich our knowledge of imagery in animal species—but four experiments (two on pigeons, and one each on monkeys and dolphins) can be quoted in support of the phenomenon of imagery being used for spatial transformations in animals.

Imagery in Pigeons

Pigeons are trained to discriminate a clock hand that revolves at a constant speed from one that moves at an apparently nonconstant velocity (Neiworth and Rilling, 1987). The clock hands are displayed on the screen of a computer monitor. In the imagery experiment, the hand disappears (e.g., after a 90° rotation or at 3:00) for some time (0.5 second) and then reappears (at 135° or 4:30). The pigeon's discriminative ability is evaluated by showing it a clock hand that speeds up during the time it is not visible. For example, the hand appears at 135° after 1 second instead of the usual interval of 0.5 second (see Figure 2.2). Results, measured by correct scores, show that the pigeon is able to estimate the position of the clock hand during its invisible journey. Further evidence that pigeons rely on an image and are not simply conditioned to time and position of the moving hand is provided by comparing their performance when the clock hands appear in novel, intermediate locations (158°) with their performance when the hands appear in trained locations and locations outside the training boundaries (202°). The pigeons remained able to make discriminations when the stimuli were presented in the novel locations, and they were able to do so correctly from the first session of testing.

These experiments suggest that the pigeon uses an image of the moving hand and that it estimates its position as a function of its velocity. In short, pigeons are shown to employ images when they are faced with this kind of task.

There is, however, an alternative to the hypothesis that the pigeon processes images: in effect, pigeons could solve the task simply by relying on the timing of the stimulus movement. Experimental control of this timing strategy was made in additional experiments (Neiworth, 1992) by altering the speed of stimulus movement, for example, by rotating the clock hand stimulus twice as fast as in the training trials. The results

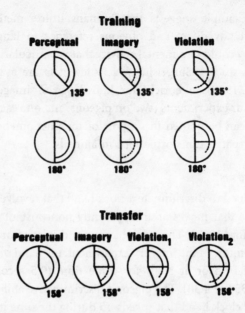

Figure 2.2. Stimuli presented in perceptual, imagery, and violation trials in the experiment with pigeons. The *solid lines* represent the position of the clock hand at various points during the experimental trials: the clock hand starts moving at 0° (12:00), disappears at 90° (3:00), and reappears or stops moving at later points (135°, 158°, or 180°). The *solid arcs* indicate the visible movement of the clock hand. The *dashed arcs* indicate where the clock hand would be had it continued to move with constant velocity. The clock hand is not visible to the subject while it is rotating within the space represented by the dashed arcs. (From Neiworth and Rilling, 1987; copyright 1987 by the American Psychological Association. Reprinted with permission.)

indicate that pigeons adjusted to the novel condition by accurately estimating the hand clock's location from processing visual information given at the beginning of each trial. A final experiment demonstrated that a dynamic visual display (versus static cues representing initial and final locations of the clock hand) was necessary for the pigeon to solve the task, thus strengthening the evidence for the imagery strategy.

Mirror-Image Discrimination of Rotated Patterns by Pigeons

An MTS task was used to train pigeons to discriminate a geometric shape from its mirror reversal (Hollard and Delius, 1982). The subject faces three response keys (see Figure 2.3). The sample stimulus is displayed on the center key. A peck at that key is immediately followed by the projection of two test stimuli on the side keys. One of the test stimuli is identical to the sample, the other is its mirror-image. A peck to the identical stimulus is rewarded with food. After long training in mirror-image discrimination, the pigeon is presented with a second task: the two test stimuli (one identical to the sample and the other its mirror-image) are both rotated to an equivalent degree (45°, 90°, 135°, or 180°) from the orientation of the sample.

Two results of this experiment are worth emphasizing: (1) the birds succeed in discriminating an image from its reversal; (2) the reaction times needed to identify the correct shape are not affected by the difference in orientation between the sample and test stimuli. In other words, pigeons solve the problem of discriminating rotated patterns but they do not seem to solve the task by means of a mental rotation strategy. In this respect they differ from human subjects tested with the same apparatus and procedure (Hollard and Delius, 1982). Data from human subjects (who "peck" with a pencil) indicate that humans do perform mental rotation.

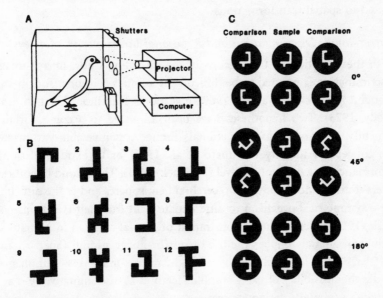

Figure 2.3. Experimental setup for mirror-image discrimination task. *(A)* Apparatus; *(B)* visual stimuli used for training; *(C)* visual stimuli used for rotation test. (Reprinted wiith permission from Hollard and Delius, 1982; copyright 1982 American Association for the Advancement of Science.)

(As noted above, human reaction times linearly increase as a function of the amplitude of the rotation.) Furthermore, the reaction times of the pigeons are systematically faster than those of humans.

Overall results indicate that the two species employ different strategies to solve the mental rotation problem. These differences may result from difference in the way mirror-images are processed in pigeons and in humans. Pigeons may distinguish both forms (a pattern and its reversal) by taking into account specific characteristics of each form. An alternative hypothesis invokes ecological factors acting on the pigeon's perception: the requirements of flight may have forced the pigeon to develop the ability to make representations of objects that are independent of their orientation; for humans, in contrast, objects are mostly characterized by their vertical orientation.

These different hypotheses need further experimental investigation. In the present state of knowledge, it can be safely concluded that the birds have elaborated a mental image of the transformed visual stimuli. However, they do not manifest image processing equivalent to that expressed by humans when they have to differentiate between stimuli that are submitted to spatial transformations.

Mirror-Image Discrimination and Rotational Invariance in Monkeys

With the exception of the pigeons in the preceding study, mirror-image discrimination is reputed to be difficult for animals (pigeons: Todrin and Blough, 1983; cats: Warren, 1969; macaques: Hamilton, Tieman, and Brody, 1973). Two hypotheses have been advanced to account for this difficulty. The first is based on animals' limited cognitive ability to master the concept of identity (D'Amato et al., 1985) or to extract relational information at a conceptual level (Premack, 1983). The second hypothesis refers to the anatomy of the two cerebral hemispheres and to the difficulty for a symmetrical organism to differentiate right from left (Corballis and Beale, 1976). A clear-cut demonstration of mental rotation in humans as well as in animals implies the use of mirror-image stimuli (Shepard and Metzler, 1971). In effect, use of left-right mirror-image test stimuli precludes the possibility of using stimulus features as discriminative cues and imposes a mental rotation strategy.

Hopkins, Fagot, and Vauclair (1993) have tested the discrimination of asymmetrical patterns and mirror-images in baboons by using the delayed matching-to-sample procedure. This work combines recent technological developments in the control of visual fixation for unilateral presentation in nonhuman primates with the mental rotation paradigm. Prior to the experiment, the monkeys were trained in the manipulation of a joystick whose movements are isomorphic with the displacement of the cursor on a computer monitor. Using this apparatus (see Figure 2.4), baboons are taught a matching-to-sample task (explained earlier in this chapter) within the constraints of a divided visual field presentation. That is, once eye fixation on the center of the monitor is assured, the sample stimulus (the letter *F* or *P*) is presented for 150 milliseconds to the right or to the left of the fixation point or, in other terms, in the left visual or right visual field. Then, two comparison stimuli are displayed on the monitor: one test stimulus matches the sample stimulus, the other is its mirror-image (see Figure 2.4). In addition, both test shapes are oriented at 60°, 120°, 180°, 240°, 300°, or 360°. Moving the cursor, by way of joystick manipulation,

Figure 2.4. Experimental setup for matching-to-sample task. *Top:* A cage is fitted with a viewport *(A)*, two hand ports *(B)*, a food dispenser *(C)*, a touch-sensitive pad *(D)*, a joystick mechanism *(E)*, and a 14-inch color monitor *(F)*. *Bottom:* The lateral display of the sample stimuli on the monitor *(left)* and the two comparison mirror-image stimuli *(right)*. (From Vauclair et al., 1993)

to the location of the matching comparison stimulus is reinforced with a food reward.

The critical feature of this procedure is the ability to present stimuli unilaterally and thus to lateralize the visual input in one cerebral hemisphere. Forms perceived in the left visual field are projected to the contralateral hemisphere, and vice versa for right visual field presentations. Since baboons are not split-brain, the visual input will very soon cross over into the contralateral hemisphere; in other words, only the initial treatment of the stimulus is lateralized.

Baboons solve the mirror-image discrimination test with an accuracy averaging 77 percent for left as well as for right visual field presentations. Moreover, the analysis of response times for right visual field presentations reveals a mental rotation curve, suggesting that the left hemisphere of the baboon but not the right utilizes a human-like mental rotation strategy. The right hemisphere of the baboon can also successfully discriminate mirror-images, but apparently the monkeys use an as yet unknown strategy that differs from the typical mental rotation phenomenon. Overall reaction times for the baboons are roughly 2.5 times faster than for human subjects tested in the same conditions (Vauclair et al., 1993). Note that in the preceding study, the two cerebral hemispheres of the human subjects seem to have equivalent rotation abilities.

Explaining the differences in response times and mental rotation rates between humans and baboons will require follow-up investigations. At present, any explanation remains speculative. It is noteworthy, however, that the letter stimuli used in this study have a specific meaning or reinforcement history for human subjects that does not exist for baboons. This fact may account for some of the differences in response times. Alternatively, one cannot not rule out other factors, such as the differences in overall size of the brain or overall differences in the sensorimotor systems or cognitive levels of processing between the two species.

These findings on mental rotation represent a significant step forward in the study of animal cognition for several reasons. First, unlike the pigeon and contrary to previous theoretical perspectives, baboons are sensitive to the effect of orientation. Second, the data suggest a common underlying cognitive basis of representational processes in humans and monkeys for this cognitive phenomenon. This shared process, which appears to exist independent of language, shows that language is not a prerequisite for representational "thought." Recently, a study of rotation invariance using an MTS procedure has been carried out in a young dolphin (Herman, Pack, and Morrel-Samuels, 1993). Although this preliminary work has not used mirror-image stimuli as comparison stimuli, analysis of differences in response times reveals that responses to 180° rotations are slower than responses to 90° rotations. This result is consistent with the activity of mental rotation. These limited data are thus in agreement with the findings in baboons, and they suggest that dolphins too can manipulate mental representations of objects.

Summary and Current Debate

All the examples presented in this chapter provide ample evidence that several behaviors manifested by animals are cognitive in nature. These manifestations satisfy the conditions and criteria for cognitive organization mentioned in Chapter 1 (flexibility, novelty, and generalization).

A big issue that will be reiterated throughout this book concerns the nature of animal cognitive processes for the animal itself. If, as the present chapter has demonstrated, animals use cognition in solving different kinds of problems, a question remains as to whether this cognition is conscious or not. Most comparative psychologists assume that the cognitive processes of animals (as well as of humans) are unconscious. For example, for Terrace (1984), as I have already mentioned, "animal cognition is not the study of animal consciousness" (p. 7). A slightly different view amounts to saying that "consciousness is simply one cognitive process among many" (Roitblat, 1987, p. 5). In this sense, consciousness refers neither to behaviors nor to external objects but to the organism's other cognitive processes. It follows from this conception that "thinking does not imply consciousness, but consciousness implies thinking. Specifically, it implies thinking about thinking" (Roitblat, 1987, p. 5).

Griffin, who is the proponent of an alternative approach to animal cognition known as "cognitive ethology" (for example, Griffin, 1992), has advocated that animals do have "subjective mental experiences," "conscious thoughts," or "subjective feelings" and that the goal of cognitive ethology is precisely to study these phenomena in animal species. The problems arising from the agenda proposed by Griffin will be exposed in due time (Chapter 8). For the time being, even if the controversy is not resolved, there seems to be a consensus among comparative psychologists that "even if one accepts that representations presuppose conscious experience, it is not the subjective quality of the experience itself that is under investigation. In fact, such subjective quality is impossible to assess: How could we possibly know what it is like to be a bat. Fortunately, the aim of a scientific study of the minds of other animals is not to find out what it is like to be a certain type of animal, but rather to clarify how mental states cause observable behavior" (Prato Previde et al., 1992, p. 86).

Let us return to the program of comparative psychology. Chapter 1 has listed some criteria (for example, the capacity for generalization) that are used to assess the cognitive nature of behaviors produced by animals in several experimental contexts. These can be further examined within the context of Piaget's theory of the development of intelligence in human infants and children and its application to animals. This developmental theory offers an agenda well suited to the study of cognitive operations as they are performed by animals. This theory and its application in the field of animal cognition is the subject of the next chapter.

3

Piagetian Studies in Animal Psychology

This chapter presents the conceptual framework of the theory of intelligence put forward by Jean Piaget. Piaget's main interests were in child psychology and genetic epistemology, and his work centered on the development of intelligence in children, the biological origins of knowledge, and the construction of scientific knowledge. The influence of his theories on research in animal cognition is not easily measured with respect to the main trends of comparative psychology presented in the previous chapter. It is obvious, however, that his works and those of his followers are related to the emergence of the cognitive sciences in the years 1940–1950 (Gardner, 1985).

One of the main interests of Piaget's theory is that it offers a suitable model for discussing many questions of animal cognition, including the problem of similarities between humans and nonhuman species (Piaget, 1971).

Developmental Psychology and Comparative Psychology

The Piagetian theory of cognitive development in humans rests on the "mutual interrelation of schemes and on the differentiation and enrichment they undergo by being constantly adjusted to the external world" (Etienne, 1973a, p. 376). Adaptation of the individual to the environment will be achieved through the interplay of assimilatory and accommodatory schemes. (Schemes are conceived of as the underlying structures and organizers of actions; they will be discussed in more detail below.) Thus, adaptation is neither the emergence of totally preprogrammed structures nor a response absolutely determined by environmental factors. It is instead a consequence of both internal factors (the schemes and their organization) and external pressures (the constraints imposed by the environment). For example, "a baby who sucks the nipple of a bottle or the corner of his blanket incorporates or *assimilates* these external elements into his sucking-scheme; in other words, he applies a pre-existing general action-pattern to these objects" (Etienne, 1973a, p. 375). Note that the first action schemes (sucking, prehension, visual fixation) find their origins in the reflexes the baby is equipped with at birth.

Thus, cognitive growth mainly consists of the differentiation, generalization, and coordination of these reflexes into organizations of actions subsumed by the different invariants just mentioned. These processes cover a large field, extending from sensorimotor accomplishments to the highest symbolic systems like mathematics. A crucial aspect of this developmental theory is its assumption that the organism plays an active role in the constructive process that characterizes the ontogeny of cognitive behaviors.

The Piagetian model has several methodological and theoretical implications for the study of animal cognition. The most important are the following:

1. The Piagetian theory offers to the scholar interested in animal cognition a conceptual framework that is both general and precise for studying nonstereotyped behaviors, whose complexity cannot easily be dealt with by more traditional models (such as stimulus-response theories of learning or the ethological concept of "fixed-action-pattern").

2. The concept of "schemes" provides an operational and experimental basis for the study of knowledge and its development, in the sense that levels and types of cognitive functions can be defined and their expressions can be observed. Moreover, schemes have a broad field of application, from object manipulation and spatial behaviors to imitation and reasoning.

3. The Piagetian approach shares a common methodology with ethology—namely, to rely, to a large extent, on spontaneous manifestations of the infant's and, later, of the child's behaviors. This method and its underlying theoretical tools stress both the functional conti-

nuity of cognitive development, through the mechanisms of assimilation and accommodation, and the structural discontinuity of cognitive growth expressed by the elaboration of the different developmental stages.

4. The concept of developmental stages (or, put another way, the ontogeny of behaviors) in combination with the idea of continuity between organismic regulations and cognitive functions offers a powerful tool for studying the similarities between human and animal cognition. In this respect, Piaget's framework is in agreement with the Darwinian view that there is a continuum of "mental faculties" between man and other species (or, put another way, a phylogeny of behaviors).

In sum, even though Piagetian theory was constructed to explain human developmental features, its biological roots and the specificity of its methods for studying nonverbal behaviors makes it a suitable framework for the comparative psychology of cognition. After all, information theories (see Chapter 1) that were originally applied to machines have been extremely useful for studying human and animal behaviors. Of course, if the full range of animals' cognitive competences is to be examined, the methods of psychological investigation must be adjusted to fit the specific perceptual, attentional, motor, and motivational characteristics of the species studied. This kind of adaptation was successfully made, for example, in rendering the object-permanence test ecologically relevant to cats (Dumas, 1992), as discussed below.

The Development of Intelligence

Given the significance of the Piagetian theory of development in animal cognition studies, I provide below a brief outline of that theory. For more detailed information, see Piaget (1950, 1952a, 1970) and Flavell (1963); and for more discussion with respect to the application of Piagetian concepts to animal behavior, see Doré and Dumas (1987).

Piaget views knowledge as a product of biological adaptation that is constructed by each individual by interacting with the environment. His model describes mental mechanisms, called "schemes," and mental operations that originate in the innate reflexes of the newborn. The notion of *scheme* is central to Piaget's theory. Schemes are the underlying structures and organizers of actions. The construction of adaptive behavior "occurs through interacting processes of assimilation of objects to action schemes, and accommodations of these schemes to the properties of assimilated objects, and through reflective abstraction of various actions on objects" (Parker, 1990, p. 19). Piaget's hypothesis rejects the idea of an absolute starting point for knowledge, as well as an absolute end point. In fact, it implies a form of continuity between biological processing and purely cognitive processing. In order to explain the transition between the domain of biology and that of cognition, Piaget's theory refers to two kinds of concepts, which are both functional and structural. The processes of assimilation and accommodation just mentioned will serve as the functional invariants between the organic and the psychological levels. The structural and discontinuous element is represented by the concept of stage. This latter concept describes sequential periods of intellectual development from infancy to adolescence.

Four periods have been distinguished in this development: the sensorimotor period (from birth to 2 years), the preoperational period (from 2 years to 6–7 years), the period of concrete operations (from 6–7 years to 11–12 years), and finally the period of formal operations (from 11–12 years on). (See Piaget, 1950; Piaget and Inhelder, 1969).

The Sensorimotor Period and Object Permanence

The sensorimotor period is partitioned into six main stages. At the beginning, reflex schemes are applied to different objects in the environment, primarily through sucking and oral contact. Reflexive behaviors lead (stage 2) to the formation of habits that concern the infant's own body (such as repeated thumb-sucking and visual pursuit of the hand). They define what is called the stage of *the primary circular reactions*. A circular reaction is the repetition of an act that usually has a pleasant effect for the infant. Between the ages of 4 and 8 months (stage 3), behaviors oriented toward objects appear *(secondary circular reactions)*. These novel behaviors are made possible by the coordination of vision and prehension. For example, a 6-month-old pulls a blanket while simultaneously watching it move or shakes an object in order to repeat an interesting sound. Stage 4 is characterized (at the end of the first year) by the application of complex coordinations of actions directed toward objects. This process begins the *tertiary circular reactions*, which can be defined as the trial-and-error manipulations of objects in relation to other objects. Tertiary reactions are almost systematic and experimental actions performed by the child; examples are the repeated dropping of objects by the child or the repetitive placing of objects in and out of containers.

The different circular reactions combine with each other, paving the way to the progressive mastery of spatial and causal relations between objects (stage 5). In stage 6, the earlier schemes will be combined mentally, so that a given scheme (for example, remove an obstacle) can be used as a means to activate another scheme (discover an hidden object). (See Table 3.1 for an overview of Piaget's developmental stages of sensorimotor intelligence and object permanence.)

Cognitive development during the first two years of life consists of the establishment of successive invariants. One of the main invariants acquired during this sensorimotor period is *object permanence,* meaning that the child conceives of objects as fixed and permanent entities. The ability to understand that an object exists even when it is hidden from view is assessed by examining the infant's reaction to objects that are moved about and hidden while the subject observes both the movement and the hiding. Object permanence is considered by Piaget to be one of the most impor-

Table 3.1. The development of sensorimotor intelligence in humans (according to Piaget).

Sensorimotor stage	Sensorimotor intelligence	Object-oriented behavior
1. Reflex (0–1 month)		
2. Primary circular reactions (1–4 months)	Acts on self	
3. Secondary circular reactions (4–8 months)	Acts on objects	Finds partially hidden objects
4. Coordination of secondary circular reactions (8–12 months)	Goal-directed sequences	Finds completely hidden objects
5. Tertiary circular reactions (12–18 months)	Discovery of new means through active experimentation	Follows sequential visible displacements
6. Mental representations (18–24 months)	Invention of new means through mental combinations	Follows sequential invisible displacements

tant achievements of intelligence. This is because the object plays the role of an invariant in all contexts of the infant's cognitive acquisitions, as, for example, in the organization of space, time, and causality.

Six steps or stages can be distinguished in the construction of object permanence (see Table 3.1). The search for an object that has disappeared from the perceptual field begins only during the second stage. The infant searches for the object only if one part is still visible or if the obstacle preventing its view is located in the immediate perceptual field. Object permanence first emerges during stage 3. Stage 4 of object permanence is characterized by the search for a completely hidden object. At this stage, the object has become permanent across time and, to a limited extent, across space. Thus, an interesting limitation characterizes this stage: if an object is hidden and discovered under cover A and then is hidden under cover B, the infant still searches for it under cover A. This behavior is known as the *typical error of stage 4*. In stage 5, successive visible displacements are mastered, and in stage 6 the infant will continue to search for the object even if it has been submitted to a succession of invisible displacements (as when the object is transported in an opaque container from A to B and from B to C in the same trial).

For Piaget, stage-6 infants are able to mentally reconstruct the trajectory of the movement of an object that is absent from their perceptual field. These stages correspond to the main stages of the development of intelligence as they have been previously described.

At the end of the sensorimotor period, the infant recognizes an object regardless of the spatial position in which it is perceived. The permanent object thus becomes the first invariant of "the practical group of displacements" whose role is to organize the child's movements in space and to structure objects' movement in the external world.

The Later Developmental Periods

After the sensorimotor period, the infant enters the preoperational period, which extends from 2 to 6 years of age. This period is characterized by the appearance of new types of behaviors, one of which is the semiotic or symbolic function (Piaget and Inhelder, 1969). This function describes the ability of 2-year-olds to represent a given object or event (called a "referent") by different means (called "signifiers"). The signifier may be a gesture, a mental image, or a word. Thanks to such symbolic tools, and in particular to language, representative and conceptual thinking (or, in other words, internalized action) is possible. These achievements are precursors to the realization of elementary inferences, or figural categorizations.

By 6–7 years, the child enters the period of concrete operations, which means that concepts (such as numbers) and principles of physical invariance (such as weight) are mastered. These cognitive abilities are endowed with important properties (mental reversibility, transitivity). The child's application of this logical framework is, however, limited to the outcome of actions on objects, in the sense that intellectual operations are still strongly linked to objects and to relations among objects.

Next comes the formal operations period (from 11 years old on), during which adolescents adopt logical strategies that are independent of their content. Thinking occurs on a hypothetical-deductive level wherein mental operations combine together and are applied to a content that has a propositional and a hypothetical status. The individual is now ready to reason on the level of merely possible events through the use of logical implications and can, for example, make "if-then" conclusions.

The ages at which children reach successive developmental stages are only approximate. They can vary according to culture, individual differences, and other factors. What is nevertheless important to keep in mind is that the sequence that goes from one period to the other (or from one stage to the next) is considered to be invariant.

Sensorimotor Activities in Animals

Object Permanence

Interest in Piaget's concepts and methods for comparative studies of animal "intelligence," and especially for comparing primate species, appears to have begun with Jolly (1964; cited in Parker, 1990), but of course Piaget himself was the first to apply his framework to animal behavior (e.g., Piaget, 1971).

Still, the first experimental work concerned the development of object permanence in domestic cats (Gruber, Girgus, and Banuazizi, 1971). This study recorded the reactions of kittens from week 1 to week 20 to the disappearance of objects. In this typical "object-permanence task," an attractive object is hidden under one or several soft cloths. The authors observed that by 4 months of age cats attain stage 4 of object permanence (a degree of object permanence that the child reaches only at 8–9 months).

Additionally, the sequence of acquisition is similar in the cat and in the human. However, the rapidity of the cat's movements prevented further investigation of its ability to master sequential visible displacements. Such studies were carried out later with one-year-old cats (Thinus-Blanc, Poucet and Chapuis, 1982), as well as with adult cats (Doré, 1986). Both studies showed that cats can search for hidden objects after visible displacements (an ability that corresponds to stage 5), but not to invisible displacements. However, stage 6 reactions have been described in 4-to-9 month-old cats (Dumas, 1992). In the typical Piagetian task, the information necessary to succeed in finding a hidden object must be processed retrospectively (the subject watches the displacement and is later required to search for the hidden object). In the adapted version administered to cats, the object is invisibly hidden while the cat is walking behind the opaque part of the detour device (see Figure 3.1). In other words, the search movement toward the vanishing object is initiated before its disappearance, a characteristic obviously more similar to the natural predatory behavior of this species. In this setup, the cat can solve the invisible displacement problem (see also Gagnon and Doré, 1994 for findings on invisible displacement test in dogs). Some forms of object permanence have also been described in other mammals such as hamsters (Thinus-Blanc and Scardigli, 1981) and dogs (Triana and Pasnak, 1981).

Domestic chicks (3 days, 6 days, and 14 days old) have been tested in a prey-object search task (Etienne, 1973b). The chick is placed between two screens on either side of a worm inside a glass tube. The worm is then pulled through the tube until it disappears behind a screen. The chick is then left for one minute in the test environment: it can get the worm either by going directly behind the correct screen or by circling first around the wrong screen and then finding its way behind the correct one. Chicks of all age groups do not immediately walk behind the correct screen. Generally, after a few trials, they orient fortuitously toward the correct screen and find the worm. From that time on, the birds learn rapidly to go behind the correct screen after the disappearance of the worm. But immediate orientation toward the correct screen emerges after only about 30 trials. Still, searching behaviors of this type do not generalize to novel spatial situations, and they do not change with age. It is noteworthy that four psittacine birds (of the parrot family) tested on the usual object-permanence task have shown stage-6 competence, demonstrating that this ability

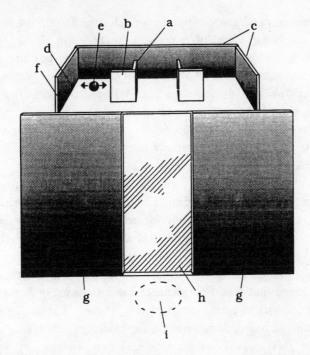

Figure 3.1. Experimental apparatus for testing object permanence in cats. Two screens *(a, b)* hide a small part of an area enclosed by a rear panel with side sections *(c)*. When experimental trials were administered, the target object *(e)* was tied to transparent strings passed through hole *d* (and the object would therefore pass behind the hiding screen); during warm-ups the object was tied to strings passed through hole *f* (and the object would therefore pass in front of the hiding screen). From a starting position at *i*, in front of a transparent panel *(h)*, the cat must walk past opaque panels *(g)* to begin its search. (From Dumas, 1992; copyright 1992 by the American Psychological Association. Reprinted with permission.)

is not limited to mammals (Pepperberg and Kozak, 1986; Pepperberg and Funk, 1990).

Object permanence has been extensively investigated in several species of nonhuman primates. All of these studies agree in their description of the sequence of development of this capacity and its similarity with human development. Species differences appear in terms of the stages attained by the animals tested, however. For instance, squirrel monkeys seem to stop at stage 4 (Vaughter, Smotherman, and Ordy, 1972); others (a capuchin or cebus monkey and a woolly monkey) reach stage 5 (Mathieu, Bouchard, Granger, and Herscovitch, 1976; see also De Blois and Novak, 1994). By contrast, chimpanzees and gorillas appear to achieve stage 6 more rapidly than the human child does (Wood, Moriarty, Gardner, and Gardner, 1980; Redshaw, 1978). The object-permanence task can be a sensitive tool for measuring fine differences between species. For example, some studies (Natale, Antinucci, and Poti, 1986; Schino, Spinozzi, and Berlinguer, 1990) have found that adult cebus monkeys, but not macaques, are able to solve the invisible-displacement task (stage 6).

It is not too surprising that animals attribute some permanence to objects according to their localization in space. The adaptive value of this ability is rather obvious, especially for predatory species that constantly monitor the position and movement of their prey. Thus, object permanence is not necessarily the result of an ontogenetic construction; it may be preprogrammed in the behavioral repertoire of a given species. This does not mean that object permanence cannot change or become more and more complex as a function of specific training (Etienne, 1984). Even so, a real change in searching strategies for hidden objects among primates appears in great apes. Does the similarity between humans' and apes' abilities imply that object permanence has a similar meaning in people and apes? Perhaps not. First of all, all studies have shown that nonhuman primates acquire these abilities earlier than human subjects. This precocity can be explained by the fact that infants of the other primate species mature earlier than human infants. For example, locomotion is functional much earlier in nonhuman than in human primates (end of the first month versus 8–9 months). As I have argued elsewhere (Vauclair, 1982, 1984; see also Spinozzi and Natale, 1986), this precocious maturation might affect the capacity to locate an object in space. In effect, one can expect that an organism that is able to move around will have to take into account the spatial relations among objects in its environment. This is not to be expected in a human infant, who will remain in a more or less stable relation to environmental objects for several months.

In addition to this feature, object permanence acquisition is contemporary with several other developments, namely the spatial and functional organization of objects (see above). Consequently, it cannot be concluded from the sole presence of object permanence that sensorimotor intelligence is identical in animals and in humans. To do that, it would also have to be demonstrated that certain other acquisitions occur, which is, as we shall see, far from being the case, even in those nonhuman primate species that attain stage 6 object permanence.

Space and Causality

The understanding of spatial relations during locomotor activities (e.g., the ability to take a short cut or to make a detour) has been investigated in a study with a 17-month-old female gorilla (Visalberghi, 1986). The experiment required the subject to locate one specific container among four (see Figure 3.2).

There are three experimental conditions, each depending on the relation between the point at which the animal is held when she sees the reward being hidden in one of the containers (observation position) and the position at which she is released (released position). Under the 360° condition, the gorilla is carried in a full circle and released at the observation position; under the 180° condition, the release position is at 180° from the observation position; and under the 90° condition, the release position is 90° from the observation position. The gorilla has no problem

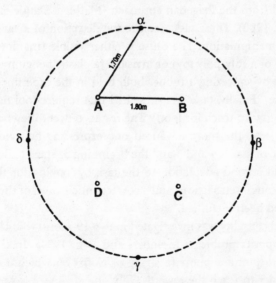

Figure 3.2. Experimental setup for test of spatial ability during locomotion. The gorilla was required to locate a specific container among four possible choices *(A, B, C, D)* after being released from observation positions α, β, γ, δ. The subject was transported by the experimenter from the observation position to the release position along the hatched line. (From Visalberghi, 1986.)

in locating the container after having been carried around in a circle and returned to the observation position (the 360° condition), but she is clearly unable to compensate for the effect of rotation under the other two conditions. When the gorilla fails, its search strategy consists of performing a systematic tour of all four containers. The latter strategy, even though it looks rather primitive (human subjects of comparable age never display this behavior: Visalberghi, 1986), nevertheless leads eventually to the correct solution of the problem; this strategy thus appears to be adaptive from the animal's point of view.

Redshaw (1978) described, in young gorillas, the progressive objectivation of space (studied, for example, through requiring the subject to find the source of sounds or make detours in order to obtain objects). This construction—namely, the ability to objectify space—includes conspecifics, objects in the environment, and the subject itself. As was true of object permanence, spatial behavior is more advanced in the gorilla than in a human infant of the same age.

The understanding of certain causal relations between objects was analyzed in two house-reared chimpanzees (aged 28 and 30 months) by using tasks derived from the Piagetian approach (Mathieu, Daudelin, Dagenais, and Décarie, 1980). These tasks require the detection of a causal relation following a manipulation. The cause is either visible (manipulation of a wind-up toy or a roly-poly toy) or invisible (a plastic penguin whose parts are activated by squeezing a rubber bulb held in the experimenter's hand under a table). Finally, tasks are presented that require tool use (use of a support in order to reach for food). The results reflect differences between the two subjects, the more advanced one expressing behaviors that are characteristic of stages 4 and 5 and the beginning of stage 6. Both subjects demonstrated a good adaptation to the task by looking for the cause of movement (causal retroaction) and a correct anticipation of the effect after the cause had been identified.

Spatialized causality was investigated in two 19-month-old chimpanzees with the "support problem" (Spinozzi and Poti, 1993). In this test, the subject has to pull a support (a string of cloth) on which a reward was placed in order to reach the reward. Only one of the subjects was able to solve the problem, while the other subject pulled on the cloth even when the reward was placed beside it. One may suppose that 19 months is the earliest age for the emergence of the skills required to solve this problem of causality in chimpanzees.

Object Manipulation

Object manipulation may be studied in animals in several ways. Experiments may be designed to elicit an animal's curiosity and reactions to novelty (Glickman and Sroges, 1966) or to reveal the morphology of motor patterns (Torigoe, 1985), or they may be conducted within the framework of play behaviors (Candland, French, and Johnson, 1978). The Piagetian approach also offers, through the distinction of different types of circular reactions, a comparative tool for assessing object-oriented behaviors. Most of the studies that have been conducted concern primates (with the exception of a study on young wolves and malamutes: Frank and Frank, 1985).

A young macaque, observed from the first week to 6 months of age, made many contacts with surrounding objects (Parker, 1977). Yet these manipulations differ from those made by human infants in terms of both their goals and complexity: most manipulations occurred in relation to the body. They thus correspond to the primary circular reactions described above; however, the reactions were generally not circular, since each of the different behaviors was rarely repeated. (The human infant, in contrast, will generally repeat pleasing actions.) This relative poverty of the manipulatory schemes of the young primate was also found in a study with a macaque and a chimpanzee (Spinozzi and Natale, 1986). Nursery-reared chimpanzees express multiple secondary circular reactions and, from the age of 5 months, tertiary circular reactions. That is, they engage in a sort of active experimentation, often with the "dropping scheme" (Mathieu and Bergeron, 1981). Finally, a study examining three 8–11-month old chimpanzees and a human infant of the same age revealed that chimpanzees indeed perform multiple manipulations with objects. In the apes, these manipulations mostly consist of simple holding or moving an object against a substrate (Vauclair and Bard, 1983). By contrast, the human infant more frequently detaches an object from the background and, furthermore, moves the object or explores its unique characteristics. This frequent extraction of objects by human infants, but not by ape infants, helps the human infant develop superior abilities for manipulation and combination of objects. The basic findings concerning the chimpanzee's relative absence of object exploration and object-object combination have been confirmed in another, more recent study (Poti and Spinozzi, 1994).

The young ape does not seem to develop during its first year the same

type of object-oriented behaviors as the human infant. The kind of manipulations seen in humans might be of evolutionary significance, because
the use of discrete and movable objects makes humans susceptible to many
kinds of arrangements and combinations, such as construction and tool
use. It must be noted that these advanced skills do eventually develop in
chimpanzees, since adults of this species have been observed to use objects
in complex ways, for example, in tool use (see Chapter 4). Prerequisite
behaviors from which later complex skills develop are evident in both ape
and human infants at 8–11 months of age, but only human infants of this
age demonstrate clear instances of complex manipulations with objects.
This human specificity will be considered further in Chapter 6, in a more
general discussion of the role of object manipulation in the development
of communicative and linguistic skills.

"Concrete Operations" in Animals

I will examine now various cognitive abilities involved in the formation
of concepts such as number, analogical reasoning, and the conservation of
physical quantities (length, volume). Because these abilities are considered
to develop in human children during the "concrete operations" period
(from the age of 6 years up), I have grouped the experimental studies of
these achievements in this section one the final stage of development, even
if their authors have not always made explicit reference to Piagetian theory.

Inferential Reasoning

A typical experimental situation for studying inferential reasoning involves
exposing the subject to a set of objects for which a given relation can be
detected ("smaller than" or "heavier than"). The subject learns, for example, that A < B and B < C (adjacent pairs) and then is tested for its ability
to understand that A < C (nonadjacent pairs). A number of studies are
available that concern different species (pigeons, rats, squirrel monkeys,
and chimpanzees). For example, Gillan (1981) trained three 5–6-year-old
chimpanzees with pairs of adjacent objects (containers of different colors
and sizes) from a series of five objects. The subject learns that E has more
food than D, D has more food than C, C more than B, and B more than
A. Then the novel, nonadjacent pair BD is presented. If the subject consistently selects the element D, it is reputed to have exhibited transitive

inference. Results (see Table 3.2) indicate that this ability was strongly present in one of the three chimpanzees and possibly in a second subject.

Inferential abilities of the same order of complexity have also been observed in adult squirrel monkeys (McGonigle and Chalmers, 1977) in experiments using pairs of objects that are, for a given comparison, always "light" or "heavy." It emerges from recent investigations with pigeons and rats that these species also show evidence of some forms of transitive inference. Pigeons (Von Fersen, Wynne, Delius, and Staddon, 1991) were trained with four pairs of visual stimuli in a series comprising five terms. The bird is rewarded for choosing A over B, B over C, C over D, and D over E. Then, when an unreinforced pair BD is presented, all pigeons choose B. Similar choices are expressed by rats (Davis, 1992) when tested on an ordered series of five olfactory stimuli. Authors differ in their interpretations of transitive inference abilities: some (Von Fersen et al., 1991; Couvillon and Bitterman, 1992) invoke a mechanism based on conditioning processes, while others (Davis, 1992) propose a spatial treatment of the task. The spatial-coding hypothesis implies that relations between objects are organized as linear representations of space marked by different spatial cues at each end of the array. Transitivity then consists of judging which member of a pair is nearer one end of the linear sequence. Experimental support for this latter hypothesis is available from further studies with rats (Roberts and Phelps, 1994).

Given the relative generality of the phenomenon of transitive inference in laboratory experiments with various animal species, it seems unlikely that all species tested so far use formal deductions to solve the task. Thus, for Wynne (1995), transitive inference performance can be accounted for

Table 3.2. Responses from three subjects in test of transitive inference in chimpanzees. Numbers indicate correct choices/total trials in presentation of four adjacent pairs and one nonadjacent pair (BD) (results for Experiment 1 from Gillan, 1981).

Subject	Adjacent pairs				BD
	AB	BC	CD	DE	
Jessie	9/9	3/8	6/9	7/10	7/12
Luvie	9/9	5/9	7/9	5/9	5/12
Sadie	8/9	9/9	6/9	9/9	12/12

in terms of simple associative learning models. But at the least it can be said that making inferences involves the integration of an ordered series of stimuli that have never been presented at the same time. (In Piagetian theory, this ability characterizes the period of "concrete operations.") This type of integration seems to be possible because the subject is able to reconstruct the series from representations of previously independent terms.

Numerical Competence

There is a clear connection between transitive inference and numerical competence if one accepts, with Piaget (1952b), that the concept of number arises from the integration of the two dimensions of cardinality and ordinality. The notion of number is mastered by the child at around 7 years of age. Before this period, the child is able to evaluate small collections of objects (an activity called "relative numerousness judgment") and to verbally designate (numeral-tag) a collection of perceptual items (a competence called "subitizing"), up to 4–6 items (see the seminal work of Gelman and Gallistel, 1978, in this area). For Davis and Pérusse (1988), "subitizing is a form of perceptual shorthand based on pattern recognition and labelling rather than formal enumeration" (p. 562).

Numerical competences of animals have been looked at in many ways, but the data are difficult to interpret because of confusion over how the number concept is defined (Davis and Pérusse, 1988). Do numerousness judgments or subitizing constitute some forms of the number concept or, alternatively, is this concept necessarily related to the performance of operations on numbers, analogous to human addition or subtraction?

Even though these questions remain unanswered, numerical competence in animals has been explored in several experimental studies (for an overview, see Boysen and Capaldi, 1993) in species as varied as birds, rodents, and primates. Three examples serve to illustrate the ways in which the number concept has been studied. Corvids can correctly select a container according to the corresponding number of dots on its lid after having been shown a set of six lids (Koehler, 1950). An African gray parrot studied by Pepperberg (1994) was able, after a proper vocal labeling of one to six objects, to enumerate subsets of objects belonging to larger collections. Rats tested in an enclosure containing six identical wooden tunnels (see Figure 3.3) are trained to enter a given tunnel (the fourth). Distances between the tunnels are changed daily, and control experiments

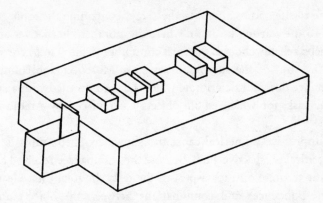

Figure 3.3. A test apparatus with a start box *(left foreground)* leading to an enclosure in which 6 identical wooden tunnels are arrayed along the left inside wall. The location of the tunnels within the enclosure was varied daily to minimize spatial cues. (From Davis and Bradford, 1986.)

are made to ensure that the rats do not use visual or olfactory cues. Rats are able to find the correct tunnel apparently on the sole basis of its ordinal position in the configuration (Davis and Bradford, 1986).

These experiments indicate that several animal species are able to estimate quantities, although interpretations of the behaviors displayed under laboratory conditions (in terms of counting or proto-counting) are far from definitive. Researchers have also looked at the abilities shown by chimpanzees to manipulate the arithmetical operation of addition (Boysen and Berntson, 1989). A chimpanzee named Sheba is trained first on a one-to-one correspondence task involving round placards with black markers affixed to them (a blank placard corresponds to "0", one marker to the number "1", two markers to "2," and three markers to "3"). One food item (then two and three items) is presented and the subject is encouraged to select the placard with one marker (or, later, with two or three markers). In a second phase, markers are replaced by Arabic numbers and Sheba is required to select the placard that corresponds to the number of food items (from 1 to 3). A test of number comprehension is then undertaken; an Arabic symbol is displayed on a video monitor and the chimpanzee is asked to select the placard bearing the corresponding number of markers. A functional counting task is finally introduced by establishing three food sites in the laboratory that Sheba visits. Food items (one to three) are placed in two of the three sites. Sheba is required to move from site to site and to attend to the arrays containing the food. She then returns to the starting position, where the number alternatives are available. Sheba reliably chooses the correct number (from 0 to 5) corresponding to the sum of the items she has seen. Moreover, the chimpanzee's performance does not deteriorate when she is confronted with a symbolic counting task, for which edible objects are replaced by Arabic number placards.

The impressive numerical capacities shown by Sheba must be viewed from a critical perspective. First, because the chimpanzee received long and systematic training on each aspect of the task, she might have learned all the correspondences and combinations between the objects and their substitutes (markers and Arabic numbers). Moreover, the numbers mastered by the chimpanzee are in a small range (0 to 5). In spite of these limitations, Sheba has spontaneously manifested interesting behaviors similar to those demonstrated by human children during counting tasks (see also Boysen et al., 1995). These behaviors are known as "tagging" and "partitioning." They include touching objects before selecting the number or moving each item apart from others in the array before making the selection. Such behaviors are equivalent to the pointing and verbalizing observed with children when they practice counting. The emergence of similar behaviors in a chimpanzee without apparent prompting from the experimenters reinforces the possibility that Sheba uses number tags that would be equivalent to a rudimentary principle of cardinality. For additional evidence of pre-counting behaviors ("summation") in the chimpanzee, see Rumbaugh, Savage-Rumbaugh, and Hegel (1987).

To finish with Sheba, this chimpanzee (along with two others) has also been tested on the transitivity task (ABCDE-ordered series of colored boxes). After the chimpanzees have learned to discriminate 4 training pairs (only the second member of each paired is food-reinforced), the novel nonadjacent BD pair is presented. All three animals succeed in choosing the correct element of the novel pair (Boysen, Berntson, and Quigley, 1993). Furthermore, Sheba generalizes her inferential skills in a task where Arabic number symbols replace the series of colored boxes.

Two additional studies investigated chimpanzees' abilities to number sets of objects in MTS tasks. In one study (Woodruff and Premack, 1981), five chimpanzees were able to match 1, 2, 3, or 4 objects with other objects representing numbers. In the other study (Matsuzawa, 1985), a 5-year-old female chimpanzee, "Ai," was trained to name the number of items (from one to six) by selecting the appropriate Arabic numerals. In a generalization test, the chimpanzee was then able to apply the correct number names to new sets of objects. Further experiments with Ai, with appropriate controls in place for size and positions of the items to be numerically named, demonstrated the chimpanzee's ability to name up to nine items and to order the numbers by reference to an ordinal scale, that is, in a sequence from the smallest to the largest number (Matsuzawa, 1991).

Analogical Reasoning

Analogical reasoning, a form of inductive reasoning, requires a capacity for judging that the relation between A and A' is equivalent to the relation between B and B'. For Premack (1976), this type of inductive reasoning requires a second-order relation. First, the subject must recognize an equivalence between A and A' and also recognize the identity between B and B'. To evaluate the analogy as such, the subject must additionally recognize an implicit equivalence of the relation of sameness in the two cases. This process has been studied with a female chimpanzee, named Sarah (Gillan, Premack, and Woodruff, 1981). Before being used in the experiment, Sarah had been trained to manipulate plastic chips standing for objects, actions, and qualifiers (among which were "same" and "different" judgments: see Chapter 6). Two sorts of configurations are presented to Sarah. In the first one (forced choice), the subject must complete an analogical relation by choosing the element B' among a set of alternative objects. In the second configuration, Sarah must pick the correct predicate (i.e., the plastic chip), same or different, by completing an analogical relation shown to her. Sarah is tested both in figural analogy problems (geometric figures related in size or color: see Figure 3.4) and in conceptual analogy problems (household objects having spatial and functional relations among them). Sarah is, for example, shown a closed lock (A), a key (A'), and a closed painted can (B) and must then choose between a can opener (B') and a paint brush (C).

For figural analogies, Sarah succeeds in 45 out of 60 trials. For functional analogies, she makes the correct selection 15 out of 18 trials. Note that in this latter test, the problem is particularly complex, given that the alternative elements (the can opener or the paint brush) may both be associated with B. In the example given above, an action can be performed on the painted can both with the can opener and with the paint brush, but the correct response is dependent on the functional relation (an opening

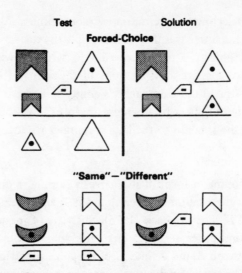

Figure 3.4. Examples of analogy problems. *Top:* a forced-choice, figural analogy problem and the solution. *Bottom:* a same-different analogy problem and the solution. In addition to the geometric figures, the experiment used Sarah's symbols for "same" (a yellow plastic rectangle with the sign =) and for "different" (a rectangle with the sign ≠). (From Gillan et al., 1981; copyright 1981 by the American Psychological Association. Reprinted with permission.)

action) between *A* and *A′*. No other species appears to have been tested on analogical reasoning tasks of the type used with Sarah.

Conservation Tasks

The concept of conservation is an important invariant acquired during the concrete-operations stage in the development of intelligence. This concept refers to a subject's capacity to judge that some properties of an object remain invariant in spite of other transformations the object may undergo. In a classical Piagetian conservation task, a subject is presented with two identical glasses containing the same amount of liquid. At approximately 5 years of age, a child is able to affirm that the two amounts of liquid are equivalent. Next, however, the content of one of the glasses is poured into a third glass, which differs in size from the first two glasses. "Non-conservers" now say that the quantities have changed, and they justify their judgment by referring, for example, to the different height of the liquids in the two glasses: when the third glass is a narrow one, such children (at 6 years of age) might declare that there is more liquid in that glass because "it's higher." After a transitional phase, "conservers" (by 7 years of age) conclude that the glasses contain identical quantities. They justify their judgment with arguments of reversibility (mental reconstitution of the pouring activity), identity ("it's always the same liquid" or "nothing has been added or subtracted"), or compensation (the level of the liquid is lower but it is wider).

This test (Woodruff, Premack, and Kennel, 1978) was proposed to Sarah, the female chimpanzee used in the experiment on analogical reasoning (see above). The conservation experiment is divided into three phases: training, test, and control. In the first phase, Sarah judges the amount of colored liquid in two jars. Three pairs of containers with different forms are used. For half of the trials, the quantity is identical in both jars; for the other half it is different. In both cases, Sarah chooses the correct response (by placing the chip for "same" or "different" on a tray between the two jars). The test of conservation is then performed in two ways: either the liquid in one of the jars is poured into a jar of different size and form, or the amount of liquid in one jar is changed (by subtracting or by adding some liquid) so that both jars have identical levels. Sarah makes the correct response under both conditions (88 percent correct for "same" and 71 percent correct for "different"). To insure that Sarah is not making judgments on the basis of a perceptual estimation but instead masters the concept of conservation, the experimenters include the control stage, during which the liquid is poured or changed out of Sarah's view. In other words, she now sees only the final level of liquids. Since her performance on this control test does not differ from chance, the overall results indicate that the chimpanzee does not rely on the mere perceptual estimation but rather takes the transformation into account. With a different testing procedure (transformation of unequal quantities of a drinkable liquid), conservation—the ability to judge in terms of "more" or "less"—was established in another chimpanzee (Muncer, 1983).

The conclusion that chimpanzees in the above experiments were able to judge that matter is conserved has been challenged on several grounds. Thomas and Walden (1985) suggest, for example, either that Sarah could have used perceptual estimation or that she was able to recall and respond to the pre-transformational state of the liquids. Because of our inability, as experimenters, to obtain direct evidence about the subject's knowledge of the pre- and post-transformational states (as psychologists may have access to judgments expressed by children through language), the evidence for conservation in nonhuman species remains scanty.

Other dimensions of conservation have also been investigated. For example, four squirrel monkeys were trained to make sameness-difference judgments of length. Specifically, the monkey was required first to indicate that two objects (rectangular yellow blocks), slightly separated but with ends congruent, were perceived as "same" and, after one of the objects was moved, to respond "same" again in reference to the now separated objects (Thomas and Peay, 1976, p. 249). Two monkeys of four succeeded in making correct length judgments and generalized their ability to different objects (green cylinders).

Summary and Current Debate

Piagetian theory serves not only as a means of testing different species with the same conceptual tool but also as a theoretical framework for reconstructing the path of the evolution of language and intelligence in early hominids. Thus, on the basis of Piagetian studies of nonhuman primates, Parker and Gibson (1979) proposed a developmental model for the appearance of language and intelligence in early hominids. The authors conclude from their attempt that "comparative data on primate development are consistent with the hypothesis that hominid intelligence evolved through a series of terminal additions of new abilities and a series of retrospective elaborations of abilities already present in rudimentary form in ancestral species" (p. 380).

Without going into the details of the model, one might criticize "Piagetian" studies of animal cognition on two principal grounds, recapitulationism and anthropomorphism. The first is specific to the application of Piaget's theory to animals, while the second is inherent to all human-oriented approaches to animal cognition.

Let us examine briefly these two questions. *Recapitulationism* refers to a theory put forward by Ernst Haeckel in the nineteenth century according to which the ontogeny (course of development) of an individual recapitulates the phylogeny (evolutionary history) of its species (see Gould, 1977, for a presentation of Haeckel's ideas and a more general discussion of ontogeny and phylogeny). The idea of recapitulationism is implied in Parker and Gibson's model, because it invokes an ontogenetic model to interpret the evolution of language and intelligence. The usefulness of this principle for this purpose can however be challenged. As far as primates are concerned, lemurs followed by macaques, cebus monkeys, and then chimpanzees would constitute a phylogenetic sequence with a corresponding increase in cognitive abilities, but in fact these species present a rather heterogeneous picture that does not follow the proposed scale of increasing cognitive performances. For example, as noted by Snowdon and French (1979), the abilities exhibited by cebus monkeys (in tool use, e.g.; see Chapter 4) surpass those of the macaques, although the cebus are more primitive phylogenetically. In the same vein, from a strict reading of Piaget's theory one might expect that counting behavior in nonhumans (an ability typical of the concrete-operations period in human children) would be found only in the great apes. But, as the section on numerical competence in this chapter has amply suggested, at least some forms of pre-counting and counting abilities may be present in rodents and birds.

The other danger in using the Piagetian framework in the analysis of animal behavior is *anthropomorphism*. This is the attribution of human-like traits to animals or, what is worse for an experimentalist, the evaluation of animal competence by exclusive reference to properly human acquisitions and performances. The risk of anthropomorphism would be increased by the naive and uncritical use of Piaget's concepts. This threat is somewhat aggravated by the fact that this theory is acknowledged to be complex. Recall that Piaget's theory points to the adult human mind as the final product of human growth and that it searches for evidence of the development of the mind in the diverse behavioral manifestations of the young child. A legitimate question must thus be asked: Is the progression of possible stages of the development of the young animal to be interpreted as steps toward the mature levels reached by the full-grown adult? Such an outcome is probably very unlikely, especially as the human cognitive structures described by Piaget, which are modeled on "logico-mathematical formalisms" (Brainerd, 1978), are far removed from the adaptive needs of animals.

These objections are serious enough to throw doubt on the validity of the Piagetian theory at large. In effect, if each species' cognitive structure is indubitably derived from its interactions with the environment, it remains that those animals with a sensorimotor organization closest to that of humans (i.e., the primates) are the best candidates for comparative purposes. Not surprisingly, as this chapter has indicated, it is precisely human and nonhuman primates that show the most similarities in terms of behavioral development. Comparing human stages with animals' capacities outside the primate order might be a more hazardous enterprise, unless care is taken to adapt the Piagetian approach to the characteristics of the species under study.

Despite these caveats, the approach has many strengths. Because of its methods, its general framework, and its comparative and developmental perspectives, the theory of Piaget is well suited to address "the general issue of the origin, nature, ontogeny, and function of animal as well as human knowledge" (Doré and Dumas, 1987, p. 230). Moreover, it is especially appropriate, given the limitations of classical models of learning, for analyzing and comparing nonstereotyped and nonconditioned activities (Parker, 1977) and for establishing a comparative developmental evolutionary psychology (CDEP). Some of the prerequisites for developing a CDEP program have been laid down in the book *"Language" and Intelligence in Monkeys and Apes* (Parker and Gibson, 1990).

4

Tool Use and Spatial and Temporal Representations

This chapter is concerned with a set of major topics in current animal cognitive psychology: the use of tools and the representation of space and time. These are considered together because the use of tools requires, in the great majority of cases, sequential organizations of actions that in turn involve a knowledge of the spatial characteristics of the objects being manipulated. Furthermore, there is experimental evidence to suggest that animals are not passive with respect to the passage of time but that they are capable of forming certain internal representations of the temporal events they experience in parallel with their processing of spatial relations.

Tool Use

The notion that animals use tools was first popularized by Köhler (1925), who observed chimpanzees piling up boxes in order to reach a supply of bananas. More recently, Goodall (1968, 1986) reported tool use in wild chimpanzees in the Gombe Stream Reserve in Tanzania. Goodall described several behaviors in chimpanzees requiring the use of an object to reach for another object. Chimpanzees use twigs to fish for termites in subterranean mounds and they use leaves as sponges to absorb water from hollows in trees. Impressive cases of tool use are not the sole prerogative of primates, however; examples have also been described in many other species of vertebrates and even in some invertebrates.

In addition to requiring sensorimotor coordinations and sets of knowledge related to the physical and spatial constraints of the objects, tool-use behaviors are usually performed within social groups and are acquired through individual practice as well as through the observation of competent adults (McGrew, 1977). Behaviors falling in the category of tool use are thus complex because they rely on several social and cognitive abilities that must be correctly performed. My decision to present these behaviors in a chapter also dealing with spatial representation was dictated by the fact that the spatial component constitutes a central aspect of tool use. In effect, tool use always involves the organization of movements in space either for the prehensile organ that manipulates the tool or for the displacement of the tool or the displacement of the subject toward the tool.

Definition and Examples of Tool Use

Given that objects can be used as tools in different contexts, any discussion of tool use must begin with a working definition. The following definition is borrowed from *Animal Tool Behavior* (Beck, 1980), which offers the most complete treatment of the subject published so far.

In order to qualify as a tool, an object, through the actions exerted on it, must be endowed with the following features (Beck, 1980, pp. 10–12):

1. The object must be detached from a substrate and must be outside the user's body.
2. The user must hold or carry the tool object when it uses it or just prior to using it and must orient it correctly in relation to the incentive.
3. Finally, the object must be used to effect a change in the form, in the position, or in the condition of another object, another organism, or the user itself.

The two examples mentioned above (chimpanzees using twigs to fish for termites and leaves to soak up water) fulfill the conditions listed by Beck to qualify a behavior as tool use. As a counter-example, consider an interesting and well-known behavior shown by Japanese macaques, sweet potato washing. Macaques will commonly dip a sweet potato in water with the probable goal of removing sand from it (Kawai, 1965). This behavior does not qualify as tool use in the sense just defined, however, because it does not fulfill the second requirement: the water in which the potato is dipped is neither carried nor held.

Continuing with the logic of this definition of tool use, *tool manufacture* is defined by Beck (1980) as any modification of an object by the user or a conspecific with the goal of improving the tool's efficiency.

There is an impressive diversity of animal tool-using behaviors, from invertebrates (the wasp) and birds (the finch) to a wide range of mammals (the squirrel, the sea otter, and nonhuman primates).

A spectacular illustration of tool use in insects was described long ago in solitary wasps (references in Beck, 1980). Females of this species excavate subterranean tubular burrows and provision them with insects that they have previously paralyzed and captured. These prey serve as food for the larvae that will develop in the burrow. Once the prey is placed and an egg deposited in the burrow, the female fills the entrance with soil and pebbles. Sometimes, the female of the solitary wasp can be observed using a pebble (or a twig or clods of earth) to pound the soil closure. This pounding or hammering behavior apparently has two functions: to make the soil more compact and to make it less conspicuous.

Several authors (e.g., Lack, 1953) have described tool use in the woodpecker finch. This bird uses twigs or cactus pines to search for larvae in holes it cannot reach with its beak. The finch holds the tool in its bill and probes into holes or crevices in bark or dead trees. Two techniques have been observed: either the tool is used to impale the insect when the bird encounters it, or it is used to provoke the exit of the insect. When these sub-goals are attained, the finch then drops the tool or holds it against a substrate with its foot and eats the insect. Sometimes, woodpecker finches holding these "spears" in their bills are seen moving from branch to branch while searching for prey.

Among small mammals, a kind of aimed throwing behavior has been observed in ground squirrels that kick sand toward a snake to make it retreat (Owings and Coss, 1977). Another well-studied case of tool use in mammals concerns the sea otter. In some populations, the sea otter opens mollusks by pounding them on a rock resting on its chest (Hall and Schaller, 1964).

Among large mammals, both wild and captive elephants use tools in different ways and for different functions (for a review, see Chevalier-Skolnikoff and Liska, 1993). Although the frequency of tool use is higher among elephants raised in captivity than among those living in the wild, it occurs in both settings, mostly in the context of body care (e.g., using vegetation held in the trunk to swat at pests).

Tool Use in Nonhuman Primates

Occurrences of tool use in nonhuman primates are numerous. At least 17 different species of primates use tools more or less regularly. This phenomenon is seen in all taxa except for prosimians (see Table 4.1). The variety of tool-use behaviors and the different contexts in which these activities have been described set primates apart from all other animal species.

New World monkeys are primarily arboreal. It is not surprising that these monkeys have been seen, in both the wild and captivity, to drop objects in situations that suggest tool use. For example, howler monkeys drop (or throw?) branches at human observers who try to follow them in the forest (Carpenter, 1934).

Among New World species, capuchin monkeys have particularly developed exploratory and manipulatory skills. Their ability to perform thumb-index opposition (often referred to as "precision grip") favors the performance of different object manipulations. Capuchin monkeys use sticks both in social context (to drive away humans, for example) and in non-social contexts (to reach for food) (Klüver, 1933). In a study of their ability to perform the "tube task" (see Figure 4.1), the monkeys are shown a horizontal Plexiglass tube containing a food reward and suitable sticks for pushing the reward out of the tube. The monkeys spontaneously use the sticks to push the food, lying in the middle of the tube, out one of the ends of the tube (Visalberghi and Trinca, 1989). They are also able to push two or more sticks end to end or actively modify a stick in order to make it more effective. The strategies exhibited by the monkeys are not always

Table 4.1. A simplified classification of living primates.

Order	Primates	
Suborder	Prosimians	Lemur, tarsier, bushbaby
Suborder	Anthropoidea	
Infraorder	New World monkeys	Marmoset, tamarin, capuchin
Infraorder	Old World Monkeys	Langur, macaque, baboon
Superfamily	Hominoidea	
Family	Hylobatidae	Gibbon, siamang
Family	Pongidae (great apes)	Orangutan, chimpanzee, gorilla
Family	Hominidae	Man

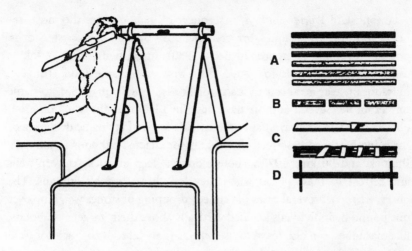

Figure 4.1. Experimental setup for the tube task. The apparatus *(left)* consists of a 30-cm transparent tube connected to a concrete base by means of inverted-V metal frames. A peanut was placed in the middle of the tube as a reward. In order to obtain the peanut, the monkey had to insert a stick in the tube and push the peanut out of the tube. Depending on the tools available *(right)*, the monkey would need to follow different strategies in order to succeed: insert and push one stick *(A)*; insert and push two short sticks, one behind the other *(B)*; modify the stick to make it thinner *(C)*; remove the attached vertical pieces from the stick before inserting it in the tube *(D)*. (From Visalberghi and Trinca, 1989.)

economical, however. For example, the capuchins will also insert very short splinters in the tube even though an appropriate stick is available. According to the authors (see also Visalberghi, 1990), although these monkeys can solve the problem (apparently by trial and error), they have a very limited understanding of the situation, and, contrary to chimpanzees, they do not acquire a mental representation of the characteristics of the tool tasks.

Capuchin monkeys will also pound and hammer different objects (tough-skinned fruits, nuts, and snakes) against tree branches (Izawa, 1979). Captive capuchins spontaneously open nuts by using a rock as a hammer (Westergaard and Fragaszy, 1987; Anderson, 1990). Reports on wild capuchin monkeys have described the use of tools by one individual in predator behaviors (oyster cracking: Fernandes, 1991) and antipredator behaviors (stick dropping or hitting on a snake: Chapman, 1986; Boinski, 1988).

Among *Old World monkeys,* macaques and baboons are the most frequent tool users. Macaques, for example, use stones to open oysters or to
pound scorpions (references in Beck, 1980). In captivity, these monkeys
can learn to use tools to obtain otherwise unavailable food: Tonkean
macaques use a metal rod to reach for honey (Anderson, 1985), to name
one example. Several authors have described baboons throwing different
objects (dust, sand, gravel) at humans. This form of agonistic dropping or
throwing has been described thoroughly for chacma baboons (Hamilton,
Buskirk, and Buskirk, 1975). These monkeys sleep on rocky canyon walls
and retreat to these places when they feel threatened by humans. The
observers report several cases of "aimed" dropping of stones weighing over
one pound by individuals situated vertically above their "targets." Baboons,
like macaques, can also learn to use objects (a metal rod) to reach for food
(Beck, 1972) or to manipulate stones to break the cement on the ground
of their enclosures, which appears "playful" (Petit and Thierry, 1993).

In addition to inanimate objects, social objects (other monkeys) may
be manipulated by monkeys in such ways that they function as tools. This
happens in agonistic encounters—interactions involving threats, physical
aggression, defense, or submission. Male Barbary macaques, for example,
will pick up and present an infant during interactions with another male
(Deag and Crook, 1971). They will even use dead infants for this purpose
(Merz, 1978). Similar cases of carrying and holding out an infant when
seeking to approach other adult males were observed in male hamadryas
baboons by Kummer (1967). This behavior may serve to inhibit the
aggressive tendencies of other males, but it can also be used to gain
proximity to females, as when a juvenile male attempts to be groomed by
the mother of the infant it has borrowed. Other instances of social tools
and their relevance to social cognition will be discussed in the next chapter.

All three species of *great apes* (in the family Pongidae; see Table 4.1)
use tools to different degrees. Although orangutans and gorillas may show
impressive tool use in captivity, the former rarely use tools in the wild, the
latter virtually never. The chimpanzee is obviously the most regular tool
user. Objects are frequently incorporated into agonistic intimidation sequences. Chimpanzees will brandish or wave branches, roots, or any available object, even a metal container (Goodall, 1971). They do so not only
during interactions with conspecifics but also when they try to intimidate
humans, to get rid of intruding baboons, or to repel predators such as
leopards (references in Beck, 1980). In the wild, chimpanzees rarely per-

form aimed throwing of objects. By contrast, in captivity there are count-less reports of chimpanzees clubbing, whipping, or hitting conspecifics or other animate objects with sticks.

The most common use of tools by wild chimpanzees is to acquire food or liquid; different techniques include leaf sponging, termite fishing, ant dipping, or nut cracking. Chimpanzees forage for termites (Goodall, 1968) by inserting a twig, a blade of grass, or a small branch into the termites' nest. Beck (1980) describes termiting by chimpanzee in the following terms: "After the termite soldiers have seized the probe with their mandibles, the chimpanzee withdraws the tool and eats the attached termites directly with its lips and teeth" (p. 85). Efficient termite fishing requires a tool of appropriate length and thickness. These constraints require the chimpanzee to prepare (manufacture) its tool in advance, as it does by detaching secondary twigs from a branch in order to make it smooth.

A very accomplished instance of tool use is that of nut cracking, reported for the chimpanzees of the Taï Forest in Ivory Cost (Struhsaker and Hunkeler, 1971; Boesch and Boesch, 1983, 1990) but also for several other populations of West African chimpanzees (see review in Boesch et al., 1994). Chimpanzees feed upon several nut species that have a protective hard shell and use clubs and stones to open them. These tool-using chimpanzees use a hard surface as an anvil (usually a rock or an emerging tree root) and a stone or a wooden club as a hammer.

Because the shell of *Panda* nuts is particularly hard to open, the chimpanzees need to select the strongest hammers and to carry them to the *Panda* trees. A study (Boesch and Boesch, 1984) on the spatial representation shown by Taï chimpanzees during hammer transport has shown that they have a precise memory of the location where potential tools are available. Moreover, the chimpanzees choose the stones while keeping the transport distance between the selected tool and the *Panda* tree at a minimum. Because visibility in the forest is limited to about 20 m, the chimpanzee cannot simultaneously perceive the tool and the nut tree. The advantage of minimizing the distance involved, as well as energy expended in transporting the hammer, seems obvious, given that transporting a 3 kg stone in one arm requires that the chimpanzee walk on three limbs. Analyses of both the weight of the tool and the distances between a *Panda* tree and the location where the tool is found, allow one to draw some inferences about the strategy utilized by the chimpanzee. It can thus be suggested that the first step is to select a *Panda* tree and then the optimal stone for transporting to that tree. These behaviors imply an elaborate representation of space, because the chimpanzee must not only be able to measure and conserve distances but also to compare several distances. Within the "cognitive map" created in this way, the chimpanzee is able to perform a permutation of objects, by changing the locations of the stones with respect to the trees, and of the reference points, by mentally measuring the distance between each tree and each place where a stone is located.

Chimpanzees as users of stones to open the nuts of oil palm were extensively studied at Bossou, Guinea, by Sugiyama and Koman (1979) and Matsuzawa (1994). During these long-term investigations of tool use in an "outdoor laboratory," three occurrences of "metatool" use (that is, using a tool on another tool) involving three different wild chimpanzees (6.5 and 10.5 years of age and an adult) were observed: the human observers witnessed, in addition to the use of two stones as a hammer and anvil, the use of a "third stone placed beneath the anvil tool as a wedge to keep the surface of the anvil stone flat and stable" (Matsuzawa, 1994, p. 361).

The Functions of Tools for Animals

It is not easy to detect species differences in tool-use behaviors, other than those concerning the frequency and the more or less different contexts in which tools are used. As the preceding examples imply, primates use tools more frequently than any other animal species. In chimpanzees, virtually all adult members of the group are able to use tools. By contrast, in a group of monkeys, these behaviors are usually expressed by certain individuals only. Comparisons of the context of tool use are less straightforward, but Beck (1980) has proposed four functional categories that may serve as a framework for comparison:

1. An animal uses a tool to extend its reach. This is the case when, for instance, a monkey uses a stick to push food out of a tube in a laboratory, or a woodpecker finch uses a twig to "spear" an insect in the crevice of a tree.
2. An animal uses a tool to amplify the mechanical force it can apply to the environment, as when chimpanzees use stones to crack hardshelled nuts.

3. An animal uses a tool in social contexts in order to reinforce the effect of an agonistic display.

4. Finally, an animal uses a tool to facilitate dealing with fluids. Several instances are described in the literature: wiping, containing, and absorbing.

According to the species and its adaptive needs, one or another of these different functions may be attributed to an animal. Only primates, however, exhibit tool-use behaviors that correspond to all four functions.

The functions listed above apply only to the manipulation of inanimate objects. Social tool use as described earlier (the use of infant chimpanzees for "agonistic buffering" by adult males) does not fit into this framework. For Beck (1980), agonistic buffering "is part of a larger set of social behaviors in which primates and probably other animals enlist the aid of others to further their interests in social interactions with third-party conspecifics. Only those in which a second party is actually carried or held conform to the present definition of tool use" (p. 66).

A Possible Hierarchy of Complexity

A classification of tool-use behaviors that differs slightly from Beck's has been proposed by Parker and Gibson (1977). This classification is based on the idea that tool use is a goal-directed form of complex object manipulation (see the distinction between different levels of object manipulation proposed by Piaget in chapter 3), involving the transformation of an object. The authors make the distinction between *true tool use* and *proto–tool use*. True tool use is the manipulation of a goal object, the tool, which is "not part of the actor's anatomical equipment and not attached to a substratum, to change the position, action, or condition of another object, either directly through the action of the tool on the object or of the object on the tool, or through action at a distance as in aimed throwing" (Parker and Gibson, 1977, pp. 624–625). Proto–tool use is the transformation of an object through "object-substrate manipulation." This classification allows us to distinguish the cognitive complexity involved in the dropping of mollusks against a sea wall by herring gulls, for example (reference in Beck, 1980), a case of proto–tool use, from the pounding of nuts by chimpanzees, which is true tool use. When a stone is used to crack a nut, the object of change (the nut) *and* the agent of change (the tool) are detached from their substrate and are manipulated.

The acquisition of tools by chimpanzees shows interesting parallels with the way human children progressively combine objects. Greenfield (1991) described a hierarchical organization in manual object combination that included three basic strategies of increasing complexity. As an example, consider the different ways of combining nested cups. In the first strategy ("pairing"), a single active object acts on a single static one (place cup A into cup B). In the second strategy, called the "pot," two or more active objects act on a single static one (place cup A into cup C and then cup B on top of C). Strategy 3, the "sub-assembly" method, involves the combination of two objects into a pair (cup A and cup B are nested), which is then manipulated as a unit in the next combination (the unit AB is combined with cup C to complete the final nesting structure).

In young chimpanzees (between 3 and 5 years of age), Matsuzawa (1994) found three stages in the mastery of nut-cracking behavior with stones (see above). The manipulation of a simple piece (holding a stone) characterizes stage 1. In stage 2, two objects are related (the pairing strategy): for example, the chimpanzees would strike a nut on the anvil by hand, without a hammer. Stage 3 is defined by the coordination of multiple actions (the "pot"): the chimpanzee places the nut on the anvil stone and then strikes the nut and anvil with a hammer stone. The three cases of metatool use described earlier might approach the strategy of sub-assembly used by human children at around 3 years of age (Greenfield, 1991).

Most instances of tool use require some forms of planning and representation. The representations involved are essentially spatial and temporal, since the achievement of the goal is often delayed in time and the goal may be invisible at the start of the behavioral sequence.

Spatial Representations

Most animal species move about in their environment, searching for food and for partners, escaping predators, or finding resting places. These kinds of movement presuppose at least a fixed reference location (e.g., "home") to which the animal must return. Except in those situations in which a landmark is constantly present either at the start of a journey or at its end, the animal is continually faced with changing information influencing its knowledge of its position and the position of objects in the environment. The representation underlying the ability to process spatial relations has been designated, following Tolman (1948), as the concept of "cognitive map" (see Chapter 2). If an animal possesses some form of spatial representation of its environment, it should be able to determine, from any position in space, any location within its familiar environment. Spatial knowledge would express itself, for example, in the animal's ability to optimize its displacements or to select novel routes.

Spatial Representation in Invertebrates

The use of spatial representations has been examined in several species of insects—wasps, ants, bees, crickets—during their journeys on familiar paths. These representations might serve to construct local maps of their surroundings. Studies carried out on ants and bees indicate that these insects can use nearby landmarks to locate a goal. The mechanism underlying these spatial abilities has been called the "snapshot model" by Cartwright and Collett (1983). According to these authors, the insect learns the location of the landmarks of a place to which it will later return by storing in its memory a two-dimensional "snapshot" (i.e., a set of images sequentially organized) of environmental cues found in that spot. "Such a snapshot does not encode the distances between landmarks or between landmarks and goal. The model shows how a bee might guide its return by continuously comparing its snapshot with its current retinal image and moving so as to reduce the discrepancy between the two" (Cartwright and Collett, 1987, p. 86). However, the correspondence between perceived images and stored images does not depend on exact identity between the two images. Thus, the digger wasp treats alike an array of pine cones and a plastic triangle marking the entrance of its nest (van Beusekom, 1948). Similarly, bees behave in a identical way in response to a solid landmark and an outline figure of this landmark (Cartwright and Collett, 1983).

The snapshot model has been elaborated into the concept of a "sequential file" memory (Beugnon and Campan, 1989), which posits the use of a sequence of temporally organized images and which allows the insect to visit different locations in its familiar domain. The question to consider now is whether these orientation systems can only use landmarks in serial order and along familiar routes or if they can also use internal representations and combinations of positions to find new routes. The capacity to devise new routes has been tested by Gould (1986) in a study with honey bees.

Bees are individually trained to follow a route from the hive to site A, located 150 m to the west, where a food source is available. After a few days, a bee is caught while setting out toward the foraging site and carried in an opaque box to a new site, B, located 160 m south of the hive. Note that site A is not visible from site B. The bee can do at least two things. Either it recognizes site B as part of a route leading to another location where food has been found in the past; in that case, it would probably be able to follow that route back to the hive. Or the bee could use visible landmarks available at B to determine the direction of site A and head directly to A even though this B–A route has never been used before. Gould found that once released, none of the bees return to the hive but that most of them set off directly toward A. The average direction chosen (324°) is rather close to the predicted direction (330°). It is worth mentioning that if the bee relies on a sequence of familiar images, it should go to the hive first and from there to site A. The fact that the animal can reach site A from site B indicates, according to Gould, that it is using a maplike cognitive representation. In such a map, the different landmarks appear to be integrated within a true topographic representation of space, allowing the bee to make use of a novel and efficient route.

The study just described appears to demonstrate that the bee constructs a representation for geometric relations between different locations in a familiar environment. Not surprisingly, this possibility has stimulated other investigators to try to replicate the original experiment. It happens that neither the bees tested by Wehner and Menzel (1990) nor those tested by Dyer (1991) take the short cut from B to A. When released at site B, either the bees return to the hive or they use their compass to set off toward the feeding site (site A) they intended to fly to at the moment of their capture at the hive.

These studies indicate an absence of map-based behavior, contradicting both the findings and the interpretation of Gould's study. Dyer (1991) nevertheless observes that bees released at B fly directly to A if they have been previously trained once on that route. The kind of information bees used during foraging was further investigated in a simulated environment (Kirchner and Braun, 1994). Honey bees were temporarily made to fly stationary in a wind tunnel on their way to a from a feeding site, 10 m south of the hive. The bees that danced indicated the direction they had flown inside the wind tunnel and not the direction of the feeder. Kirchner and Braun concluded that these results weaken the map hypothesis, because the bees appeared to use the distance and direction information acquired en route and not the landmarks surrounding the hive and the feeder to find the food source.

To conclude, the question as to whether bees have cognitive maps is highly controversial. A bee's reference to a cognitive map might, in fact, depend on context (e.g., the distance between different locations, the availability of different kinds of landmarks, and the use of different orientation mechanisms). If in some conditions the bee orients in space predominantly by relying upon topographical information, it does not seem to use a topographical map in the sense vertebrates make use of map-based information (see below). Rather, its orientation system relies mostly on stacks of snapshots activated in sequences as the insect flies toward its feeding station. Nevertheless, this system shows some flexibility in the use of the different representations of the landscape that the bee has memorized and in which it moves.

Spatial Representation in Vertebrates

The fact that animals live in a spatially heterogeneous environment clearly indicates that they have perceptual and representational mechanisms for obtaining, storing and processing information about the environment, even if only to recognize landmarks in it (Roitblat and Von Fersen, 1992). Several animal species appear to maintain rich representations of their environment. For example, Clark's nutcrackers harvest pine seeds during the late summer and hide them in thousands of discrete subterranean caches. These birds apparently recover their own caches by remembering where they are stored (Balda and Kamil, 1989). Laboratory research with nutcrackers is feasible, as illustrated by a study by Kamil and Balda (1985). In this experiment, birds had to store food in cache sites selected by the experimenters (see Figure 4.2) and then find the caches during the recovery sessions. For example, they would have a choice of 18 available cache sites and 10 days later would have to recover the food. Their recovery performance was above chance. These findings help elucidate such mechanisms as site preference or path selection and "strongly support the hypothesis that cache recovery involves spatial memory in nutcrackers" (Kamil and Balda, 1985, p. 108). Other birds, such as marsh tits and chickadees (Shettleworth, 1983), also cache food. The latter can remember caches for up to at least 28 days.

These abilities are usually discussed within the framework of the "cognitive map" (see Chapter 1 for definition), but the concept of the cognitive map does not fully explain animals' navigational abilities. "Cognitive mapping is by no means a complete explanation of how animals are able to get around in the world. It is largely a metaphorical statement about what

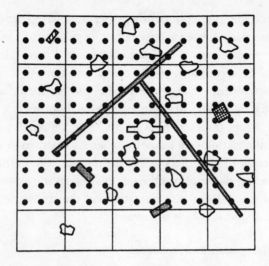

Figure 4.2. Diagram of the experimental room (3.4 m by 3.4 m) used to test spatial memory in nutcrackers. The experimenters hid food in 18 of the 180 possible caches in the floor. The room also contains rocks, logs, 2 perches, and a centrally located feeder. (From Kamil and Balda, 1985; copyright 1985 by the American Psychological Association. Reprinted with permission.)

sort of information they collect and how they organize it" (Menzel, 1978, p. 377; see also Vauclair, 1987).

The Chimpanzee and the "Traveling Salesman Problem"

One chimpanzee from a social group of six is carried about a large enclosure (4,000 m^2) by an experimenter. This experimenter, carrying the chimpanzee on his shoulders, is accompanied by a second experimenter, who hides one piece of food in each of 18 randomly selected locations. The chimpanzee watches the baiting procedure, which lasts about 10 minutes, and is then returned to the group. The experimenters leave the field and release all six animals simultaneously two minutes later. Time, the identity of the animal shown the hiding places, and the paths taken to visit the food caches are recorded.

The results (Menzel, 1973) indicate that, overall, the knowledgable subject found about 13 caches per trial. When the routes followed by the chimpanzees are analyzed, it becomes obvious that the animals do not simply follow the route along which they were carried; they use instead a route that is far more efficient than one would expect from chance (see

Figure 4.3). Menzel interprets the data to mean that each subject proceeds according to a least-distance principle. This principle, also known as "the traveling salesman problem," can be expressed as follows: "Do as well as you can from wherever you are." The problem is to minimize the distances between the different locations. In brief, the chimpanzees appear to perceive directly the relative positions of selected classes of objects and their own positions within this scale of reference.

"Especially since a novel set of locations was used on every trial, and the informed animal ran directly to hiding places that were not visible from the point from which their run commenced, the term 'cognitive mapping' seems as good a descriptor as any" (Menzel, 1987, p. 66). In brief, chimpanzees' "achievements are a good first approximation of those

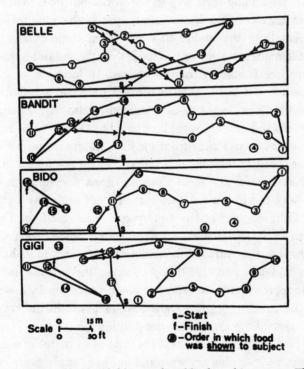

Figure 4.3. Location of hidden food items found by four chimpanzees. The connecting line traces the order in which the various places were searched and gives a rough idea of the animal's general travel routes. (Reprinted with permission from Menzel, 1973; copyright 1973 American Association for the Advancement of Science.)

at which an applied scientist would arrive from his real maps, algorithms and a priori criteria of efficiency" (Menzel, 1978, p. 407). Menzel's findings about chimpanzees' knowledge of the space within their compound nicely complement the work on spatial abilities exhibited by the chimpanzees under conditions of poor visibility in the Taï Forest, for which a least-distance principle has been hypothesized to rule the chimpanzees' decisions in transporting hammer stones to nut trees (see Boesch and Boesch, 1984, and the preceding section on tool use). Similar spatial-memory abilities were described in zoo experiments with yellow-nosed monkeys (MacDonald and Wilkie, 1990) and with gorillas (MacDonald, 1994).

Spatial Representation in Bees and Chimpanzees

Experiments show that both bees and chimpanzees rely on "spatial representations" for finding their way in the world, but does that mean that bees and chimpanzees "think alike" about space? What kinds of differences are apparent from the studies that have been completed so far? One obvious difference between the two species' spatial abilities is quantitative: the chimpanzee is able to connect at least 18 locations in space with original routes, the bee is able to connect only two positions with an original path. There seems to be a difference also in the range of the species' spatial abilities. When the bees tested by Gould (1986) are released several hundred meters away from the hive, in an area probably located outside the limits of their familiar range, they do not return to the hive; they get lost. The fact that Menzel's chimpanzees were tested in a familiar enclosure and not in a novel environment does not allow a strict comparison of the abilities of the two species. The results of the control experiment by Gould (see above) indicates nonetheless that the spatial map of the insect is restricted to certain specific zones in its home range (Beugnon and Campan, 1989). By contrast, the primate may use an allocentric representative spatial system—in other words, a system that includes the structuring of any given set of places in the environment organized around any given reference position.

It is important to remember that the concept of cognitive maps is used in a descriptive, metaphorical sense and not in an explanatory one. Thus, our present knowledge about the way animals build and use their "maps" is still preliminary. Nevertheless, some properties of the map can now be emphasized. For example, a map functions locally (which is the case for insects in their home range), but it may also function more globally (which is the case for chimpanzees able to integrate several places and their relations. In fact, we now have several tools with which to evaluate the topographical nature of representations in different species; experimenters may, for example, design tasks that make their subjects plan novel routes or take detours and short cuts to study how representations function in different species.

Reactions to Novel Objects

One method for investigating the spatial representations an animal establishes consists of analyzing its reactions to novelty. Such reactions will be examined in two species: first, in hamsters presented with a rearranged configuration of objects in an open field and, second, in baboons exposed to novel objects in their outdoor enclosure. By examining which kinds of change elicit a renewal of exploration and which kinds of exploration are performed, investigators have established valid indicators of the detection of change and of the animal's ability to respond to it, or, in other words, to "map out its environment."

The experiments with hamsters are based on the habituation procedure. It has been found that exploratory activity decreases over time (the hamster becomes habituated to its environment) and that it is reactivated after a change occurs in the environment. If a spatial change induces a renewal of exploration, as measured, for example, by the number of contacts made by the animal with an object placed in the experimental environment, then the detection of the change by the animal can be interpreted as an indication of the nature of its "internal model" and the kinds of spatial information coded in it. An experiment is carried out with hamsters in a circular open field (105 cm in diameter) to evaluate their reactions (number of contacts and investigation time) to the rearrangement of four familiar objects laid out to form a square (Poucet et al., 1986; Thinus-Blanc et al., 1987).

After two 15-minute exploratory sessions, four groups of hamsters are subjected (8–10 hours later) to experimental conditions in which different classes of spatial information are manipulated; for example, a change might be made in the topological relations between objects or in their overall geometric structure. The hamsters do not react selectively to a single displaced object in the configuration, but they react to all objects when the arrangement is entirely new. They reexplore novel geometrical configurations, but these reactions disappear with nongeometric changes (as when the distances between the objects are modified). Altogether, these results indicate that the animals form a cognitive map of their surroundings through exploration. This representation is called upon later to adjust to changes in certain spatial parameters within the external arrangement of objects defining that map.

Reactions to novel objects were studied in a group of 14 Guinea baboons living in an outdoor enclosure of 640 m². The test environment contains a dead tree, rocks of various sizes, and a wooden construction on which the baboons can climb. The procedure consists of placing each day, for twelve days, a new object in different locations in the enclosure while the monkeys are absent (confined indoors). Objects are either natural (e.g., half a coconut shell) or artificial (e.g., an aluminum mug). The baboons are thus presented each day with a novel object and, from the second day on, to objects from the previous days. The troop as a whole demonstrates an excellent ability to react rapidly to the new objects: 11 out of the 12 new objects are discovered within 3 minutes of their presentation. Furthermore, the novel objects are the first to be approached (Joubert and Vauclair, 1986).

It seems clear that the baboons' knowledge of the environment, and of changes occurring in it, falls into the category of behaviors described as "cognitive mapping." Further existence of a mapping of the environment by the group of baboons as a whole is provided by the following facts: (1) one single presentation of the object is sufficient for each animal to memorize it and to categorize it as being known; (2) new objects are mostly manipulated (touched, grasped, and sometimes transported), whereas old objects are explored more (by simple visual fixation or sniffing) than manipulated, suggesting that a distant check is enough to ascertain their identity.

The baboons' performances are in this respect similar to the reactions of a family group of marmosets (living in a large greenhouse), which were able to select a single object among up to 30 simultaneously presented test objects (Menzel and Menzel, 1979).

Structured Environments for Testing Spatial Representations

Mazes have played an important role in the development of experimental animal psychology: At the beginning of the twentieth century, they were the favorite apparatus of psychologists for testing animals (rodents, cats, dogs, and primates). As observed by Olton (1977), if one accepts that the main element of problem solving in laboratory conditions is for the animal to discover what is required by the experimenter, then the maze technique is particularly well adapted for laboratory testing, because in this case the instruction is inherent to the apparatus itself. I will consider two laboratory situations with rats in experimental settings well suited to the examination of spatial representation. The first involves the use of a radial maze to study the role of external landmarks in the spatial behavior of the rat. The second is aimed at testing the kind of spatial encoding used by birds and rats when no extra-maze cues are available.

The radial maze and the configuration of external cues. The radial maze is a spatial maze made of 8 arms radiating from a central platform. Each arm is 60 cm long and 10 cm wide and has a small cup at the end in which food can be placed. The rat is placed at the center of the platform and then required to go to all the arms for the food rewards without repeating any arm visit (Olton and Samuelson, 1976). Rats perform well in this maze: they visit an average of 7.9 arms during their first 8 choices. Controls to eliminate the possible use of intra-maze cues (related to the food, odor trails, etc.) have led to the hypothesis that the rat uses a spatial strategy based on the formation of a list of places previously selected and that should not be chosen again. It is likely that this list and the location of each arm is made on the basis of extra-maze cues.

Two main hypotheses might explain rats' performances in the radial maze. Hypothesis 1 invokes the notion of "working memory" (Olton, 1978), which refers to a storing of the list of the places that have been previously chosen and that should not be repeated. According to this view, the location is considered as an item in a list that can be processed with some degree of independence from its spatial reference. Hypothesis 2 invokes Tolman's concept of the cognitive map, according to which the rat forms a knowledge about the different topographical relationships in the maze that guides the behavior of the rat.

These two conceptions of spatial performance have been studied in several experiments using the radial maze (Suzuki, Augerinos, and Black, 1980: see Figure 4.4). The first experiment confirms previous findings that rats can indeed utilize extra-maze cues to locate the different arms. A second experiment is designed to assess the way these cues are used by the rat: either each arm is used independently of other arms (hypothesis 1), or rats use stimuli in a configurational manner (hypothesis 2).

Rats are first forced to visit three preselected arms of the radial maze;

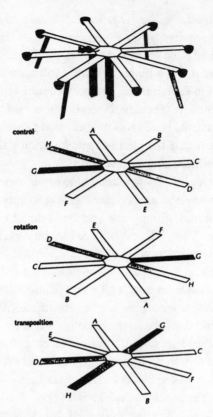

Figure 4.4. Experimental setup for the maze test. Food was left in a cup at the end of each arm. Seven visual cues (identified by letters) were attached to the wall in front of the arms. Each trial included three phases. In phase 1, the rat made 3 forced visits of 3 preselected arms (doors prevented access to the other arms). In phase 2, during the rat's confinement, two kinds of cue manipulations were performed: a rotation and a transposition. Moreover, for control subjects the cues were rotated 360° and brought back to their original position. In phase 3, the rats could visit the 5 arms that had had previously been closed. Trials were presented in the following sequence: a control trial, the first test trial, 3 control trials, the second test trial, and 3 control trials.

the rat is then confined to the center of the platform for 2.5 min. During this confinement, the five previously unchosen arms are baited (the remaining arms are left unbaited), and the cues surrounding the maze are subjected to two kinds of spatial transformations: the entire landmark array may be rotated 180°, which means that the cues keep their spatial relations but are differently located with respect to other possible cues in the environment, such as sounds or odors; or the cues may be transposed, which has the effect of altering the visual configuration of cues. If the rats behave according to the list model (hypothesis 1)—that is, each arm and its corresponding cue are treated separately—then neither transposition nor rotation should affect their performances. If, by contrast, the rats use a maplike system to locate the arms (hypothesis 2), then performance should decrease if the cues are transposed but not if the array is rotated.

The results (see Figure 4.5) indicate a significantly greater decrease in correct choices after transposition of the cues than after rotation of the array. In the latter case, performance is only slightly reduced from the control situation, in which the cues undergo no change. The authors conclude in support of the second hypothesis: since the different arms forming the maze are *not* treated separately by the rat, transposing the cues indicating relations among the arms disrupts the rat's performance. In this case, the rats are able to locate the arms by using the configuration of the maze and the topographical relations between objects. Such relations are the ingredients of cognitive maps.

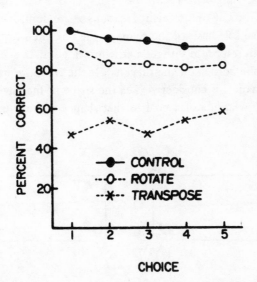

Figure 4.5. Results of the radial maze experiment: mean percentage of correct choices plotted against the ordinal number of free choice after the control trial and after manipulation (rotation or transposition) of cues. (From Suzuki et al., 1980.)

Environmental cues. As a complement to maze experiments, investigations of mapping abilities may instead be adapted to exert more precise control over the kinds of spatial features that are used by animals to orient themselves toward places and objects of interest. A limited but suggestive number of experiments with rats and birds have focused on place-finding behavior within a familiar environment.

The principle of these experiments has been elaborated by Cheng (1986) in a study with rats. A rat is placed in a rectangular box (120 cm by 60 cm), whose space is defined both by the geometric arrangement of surfaces (the rectangular form of the box) and by locational information provided by conspicuous cues (visual, tactile, and olfactory) attached to each of the four corners of the rectangle (see Figure 4.6). In the original experiment, no extra-maze cues are available to the rat. During testing, the rat is shown a dish of food placed in a randomly selected corner of the rectangular box. When the rat has eaten food several times at the same location, it is removed from the box and placed, 75 seconds later, in a second box, strictly identical to the first box with food buried in the same place. The second box is used to prevent the animal from using olfactory marks it might leave on the food. Any digging at a localized area is considered evidence of the rat's searching behavior. At the beginning of a trial, the rat is placed in the second box at one of eight locations at random. If spatial coding operates on the sole basis of the shape of the box, then two locations are equivalent with respect to this geometric information: the correct location and its opposite corner on the diagonal. If the coding implies that geometric information is considered (i.e., the shape of the environment and the geometric relations of a goal to that shape) without using the local

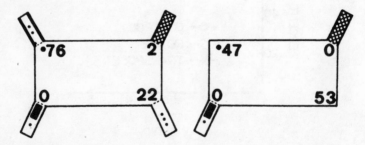

Figure 4.6. Experimental setup and results for test of spatial modularity in rats. The numbers indicate the percentages of digs at each corner. The correct location of the buried food is indicated by the filled circle. (From Cheng, 1986.)

cues placed at the four corners, then subjects should produce 50 percent correct digging responses and 50 percent rotational errors, that is, digging in the location in the same geometric relation to the shape of the box.

In one experiment, a novel location is selected for each trial: the rats make almost as many rotational errors (31 percent on average) as they make correct choices (47 percent on average). Thus, the rats have made relatively little use of the local cues to find geometrically equivalent places. The distribution of their responses speaks in favor of the preponderance of the macroscopic structure of the environment for establishing the subject's orientation (Cheng, 1986). In another experiment (see left side of Figure 4.6), food remains at the same location from trial to trial: on average 76 percent of the digs are directed to the correct location, 22 percent on average are rotational errors. The reduction of errors in this situation indicates that the rats use, at least partially, information provided by nongeometric features. In a third experiment (see right side of Figure 4.6), cues at the correct corner and at the corner diagonally opposite to the correct corner are removed. Only the locale cues associated with the corners adjacent to the correct corner are maintained. In this situation, rats choose the correct corner and its opposite diagonal at almost comparable proportions (47 percent and 53 percent, respectively).

Cheng proposes a modular conception (after Fodor, 1983) of the rat's spatial representation in the conditions tested. The rat perceives its space through the use of a geometric module, independently of the overall arrangement of nongeometric features. These latter features are stored in a different module, a featural subsystem. This second module is for distinguishing two locations that are geometrically equivalent. In this case (as indicated by the results of the third experiment), only the locale features near a geometric address are used and "glued" to the geometric module.

A replication of Cheng's experiments has helped to clarify the conditions under which rats more or less strongly rely on the macroscopic shape of the environment in a place-finding task (Margules and Gallistel, 1988). In short, when the lighting conditions allow for the use of extra-maze cues, these cues are used (see the analyses of spatial behavior in the radial maze above) and rotational errors drop to an average of only 4 percent. However, when extra-maze cues are not visible, rats use geometric information rather than the distinctive sensory features of the surfaces that define the shape (on average: 35 percent correct choices, 31 percent rotational errors, and 34 percent misses). Finally, results from chicks tested with the same rectangular box (Vallortigara, Zanforlin, and Pasti, 1990) converge with Cheng's findings. The experiment with chicks provides additional information on the interaction between the two modules. The first spatial information to be memorized concerns the overall arrangement of distant landmarks (here the rectangular shape of the cage); then, the locale features near the food are coded. In addition, the fact that geometric coding is used by the birds in a situation where it is not strictly required—namely, when there is no ambiguity with respect to the use of locale cues—indicates that spatial representations in chicks are hierarchically organized, with the geometric module dominating featural information.

Levels of Abstraction in Cognitive Mapping

The use of a map by animals to find a position in the world requires establishing a correspondence between what is currently perceived and what is preserved on the map (the "memory image"). An obvious question is the level of abstraction at which this comparison operation is carried out. (Gallistel, 1989, p. 170)

As described above, insects such as bees seem to be able to adjust to partial correspondences between the current image provided by perception and the memory image. The limits of these matching operations have also been investigated in studies of spatial memory with nonhuman primates.

"Map reading" by chimpanzees. Four chimpanzees were tested in an outdoor enclosure (1,950 m²) by Menzel, Premack, and Woodruff, (1978). Because chimpanzees become emotionally upset when they are separated from their preferred companions, the subjects are tested in pairs. The task for the subject is simply to find a familiar caretaker who has walked out into the outdoor enclosure and disappeared from sight. The enclosure contains several hills, tree stumps and a variety of large objects, which means that a person could easily escape detection simply sitting or lying down.

The pairs of chimpanzees are tested in different conditions. (1) Both subjects get a direct view (through a window) of the familiar caretaker entering the enclosure; the caretaker shows the subjects a piece of food and then proceeds to a predesignated hiding place; the subjects are released about 90 seconds after the disappearance of the caretaker. (2) Only one chimpanzee gets a direct view of the caretaker in the enclosure. (3) Both subjects watch the caretaker in the enclosure on a small (23 by 18 cm) black-and-white TV monitor; the picture is taken from approximately the same perspective as the window, but since the screen covers only about a 50° angle and lacks color, it provides rather impoverished locational information by comparison with a direct view (for example, three-dimensionality, life-size image, and motion parallax are all absent). (4) Control trials are also run in which no information at all is provided.

The results show that the chimpanzees perform differently as a function of the experimental conditions. When both subjects (or only one) get a direct view, after the first 20 trials they are almost perfect at finding the hidden caretaker. But more significantly, the rate of success for the TV-viewing chimpanzees is intermediate between the rates for direct-viewing chimpanzees and the controls. In fact, the chimpanzees exposed to the TV image behave as if they are getting a direct but impoverished view of the caretaker's movements.

Most discoveries by the direct-viewing chimpanzees and many by the TV-viewing chimpanzees occur relatively quickly after release. It is also obvious that watching TV provides locational information superior to the information available in control conditions. In sum, this experiment shows that juvenile chimpanzees can, with little training, match what they see on a screen to what is "out there" in the real world, and thereby determine the relative location of a hidden goal object in a familiar outdoor field. In this sense, the information provided by televised images could be somewhat equivalent to "reading" information from a map.

Manipulation of visual information available to baboons. Guinea baboons, members of a social group living in an enclosure (650 m²), were tested for their memory to locate objects hidden in space under varying conditions of available information (Vauclair, 1989, 1990a; this group's reaction to novel objects was also studied, as reported above). Their compound is connected via a tunnel to an indoor housing area. The enclosure contains a wooden climbing construction. Stones are scattered on the ground and serve as hiding places for hazelnuts in experimental trials.

A trial starts when an experimenter enters the enclosure while the subject is restrained in a section of the barred tunnel outside the enclosure. The experimenter walks directly to a randomly selected location and stops in front of it. He then shakes the bag containing the hazelnuts to attract the subject's attention and slowly hides a hazelnut under a stone. He then leaves the compound and releases the subject (on average after 90 seconds) into the enclosure to find the food.

Upon finding the food object (the monkey turns over or moves the stone and eats the hazelnut), the subject is returned to the tunnel for the next trial.

This procedure differs from Menzel's (1973) work on spatial memory in the chimpanzees (see above). In Menzel's experiment the subject is carried around to each hiding place in the enclosure, whereas in the present experiment the location of the hidden object is viewed at distance (between 15 and 45 meters, depending on the location selected to hide the nut).

Two male Guinea baboons (7 years of age) were tested. Two viewing conditions are used in a first experiment. (1) The subject has a full view of the enclosure through a window (40 cm by 16 cm) in the barred tunnel (binocular vision). (2) Bars overlooking the enclosure are replaced by a small central hole (1 cm in diameter), requiring the subject to peer out with one eye (monocular vision). The reason for comparing usual binocular viewing conditions with monocular conditions and their effects on search accuracy was to investigate a proposition made by Gibson: "when it is carefully arranged that a picture is seen through an aperture so that the frame is invisible, the head is motionless, and only one eye is used the resulting perception may lose its representational character" (cited in Menzel, 1978, p. 415).

The two subjects are tested in the same number of trials in each condition, with binocular trials alternating with monocular trials. On more than half of the trials, subjects go directly to the correct location in both conditions. In addition, they perform at a more than 80 percent correct after only one error (i.e., turning over a wrong stone) in both conditions. It can thus be concluded that restricting vision to monocular cues has no significant effect on search accuracy by the baboon.

In a second experiment, a color video monitor (36 cm in diameter) is placed in front of the same window through which subjects viewed the experimenter in binocular trials. Subjects are tested under two conditions: either the monitor provides a picture of the enclosure corresponding to the familiar view (TV1), or the camera is located in a tree (at a height of 4 m) and at 15 m from its initial position (TV2). When the camera is in the tree, the monitor displays a kind of "aerial view" of the enclosure displaced about 45° from the original view. Although the subjects had never been previously exposed to TV monitors, they adapted to the video conditions in the sense that the image provided by the monitor prompted searching. This phenomenon is illustrated by the following two results for TV1: (1) trials without error appear from the start of testing—in fact, one-third of the trials are concluded without error (see Figure 4.7); (2) a significant proportion of trials (90 percent for one subject and 82 percent for the other subject) cannot be explained by a random search. Overall performances during TV2 trials deteriorate compared with the results for TV1 trials, but more than one-third of the searches are still made without error (38 percent and 34 percent for the two subjects). Similar proportions of the searches seem to be randomly determined, but qualitative analyses of search patterns reveal that several unsuccessful searches are not random. For example, one baboon visited the two rows in the far right section of the enclosure during a trial in which the correct location was in the last row in the back. In this particular case, one can hypothesize that the confusion is due to a lack of compensation for rotation in matching the image in TV2 with the usual view of the compound.

To summarize, it can be said that the restriction of peripheral vision imposed by monocular viewing did not alter the two baboons' capacity to remember and locate various hidden food stores. Moreover, the video experiments show that bidimensional, reduced images can serve as representations of the real world. Finally, the video technique may be an "ecologically" valid tool for studying the transition from vision to picture perception in the baboon.

At a more general level, it can be concluded that chimpanzees and

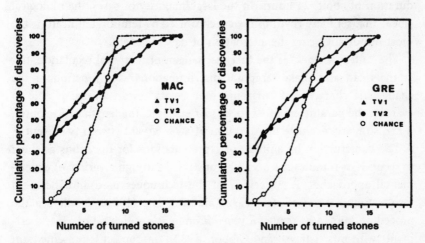

Figure 4.7. Results for test of spatial memory in two baboons viewing enclosure via a color video monitor under the two viewing conditions (TV1 and TV2). The percentage of trials in which discoveries were made after a number of stones were overturned (e.g., 1 = nut found on first try; 5 = nut found on fifth try) is plotted cumulatively.

baboons may be capable of using maps in the literal sense as well as in the metaphorical sense. Further experiments might explore the extent to which they can perform effectively when the image is rendered more and more abstract, symbolic and non-iconic, or when it is displaced in space and time from the actual event that is represented. For example, video pictures showing systematic deviations from the original perspective (e.g., rotation of the video image by 180°) could be used, or playback experiments could be conducted, to investigate the temporal and spatial determinants of visual information processing.

Temporal Representations

In addition to coping with spatial demands, animals have to manage their time. The basis of time processing is a biological clock that generates rhythms which are themselves related to the functioning of endogenous oscillators (Gallistel, 1990).

The behaviors of most animal species usually follow a regular periodicity, which can be of a short duration, as for the wingbeats of a bird, or of longer durations, such as those associated with feeding, locomotor behaviors, or seasonal variations. The most common of these rhythms is the daily or circadian rhythm, which has, as suggested by its name, a duration of about 24 hours. In the large majority of cases, the biological clocks that set these rhythms are synchronized by temporal indicators, the most important being the alternation of day and night.

The "internal clock" is the first mechanism of temporal regulation, but animals can also make temporal discriminations as a function of the changeable character of environmental events. As an animal is able to evaluate the parameters of its space (for example, the length of its route), it can similarly evaluate the duration of an interval between two events.

The construction by animals of representations for durations as short as seconds or minutes has been demonstrated through a variety of experimental approaches. For a review of the techniques used and the main results found, see Gallistel (1990) and Roiblatt (1987). I will use the procedure known as *temporal generalization* as an illustration. In experiments with rats (Church and Gibbon, 1982), the subject is presented, on each trial, with a temporal stimulus (a noise of a given duration). A specific duration is defined (called the "reinforced duration"), at the end of which the rat can press a lever to obtain some food. If, however, the duration of the temporal stimulus is longer or shorter than the reference duration, then the rat is not rewarded with food. The rats quickly adjust to the situation, by decreasing their lever presses for durations that are either too short or too long and by increasing their responses to the reinforced duration. In brief, the brain of the rat can represent elapsed temporal intervals and can compare the magnitude of the currently elapsed interval with a temporal standard stored in memory.

Several models have been proposed to explain how organisms bring into correspondence the representation of past events with the course of the presently experienced events. For example, an information-processing model (e.g., Gibbon and Church, 1984) has been proposed to explain the data obtained with rats. In this model, the rat has access to an internal clock, a working memory, a reference memory and a "comparator." The internal clock is made of a pacemaker, a switch, and an accumulator. The pacemaker provides pulses that are sent to the accumulator when the switch is "on" (that is, when a signal is emitted in the environment). The number of counts or units in the accumulator can be transferred either to reference or to working memory. In principle, reference memory will store past experiences, such as the reference duration in the experiment reported above. When the values are stored in working memory, as time elapses, these values or those of the accumulator are compared by the comparator with the values stored in reference memory. When the value in the accumulator is close to the value in reference memory, then the comparator sends a "yes" or "no" message enabling or suppressing the production of a response.

Killeen and Fetterman (1988) proposed a behavioral theory of timing as an extension and refinement of the preceding model. (See also Richelle and Lejeune, 1980, on the role of behavior as a mediator of temporal regulations.)

Summary and Current Debate

Behaviors that display an animal's abilities to move appropriately through space and use tools have been considered throughout this chapter as hallmarks of the cognitive achievements of animals. This chapter has also suggested that the use of a tool is not an isolated phenomenon because it solicits the mediation of several cognitive abilities, such as mastery of spatial relations. The relation between tool use and other cognitive abilities can be further extended by considering how, in humans, the use of tools is integrated with other acquisitions, such as language.

Several theories of cognitive development (such as the ideas of Piaget presented in Chapter 3; but see also Bates, 1979, and Greenfield, 1991) assume that the use of tools and language are intricately related and imply similar cognitive abilities. The relation between these two activities, with respect, for example, to the presence of partially similar syntactical rules, has also been stressed by linguists (Lieberman, 1975). In the same line of thinking, neuroscientists (Kimura, 1979) have postulated common neural controls for speech and manipulatory activities. Thus, the topic of tool use is at the interface of research on several cognitive skills, including social learning, the control of gestures, imitation, and so on. For a thorough review of these questions, see *Tools, Language and Cognition in Human Evolution,* edited by Gibson and Ingold (1993). For example, one important question concerns the relation between imitation and the acquisition of tool-using skills by an animal. Although this problem will be considered in some detail later (see Chapter 7), it can already be said that, with one possible exception, animals (including nonhuman primates) do not seem to imitate their conspecifics in learning how to use tools.

There are some other interesting ways of pursuing the discussion of tool use in animals within the perspective of comparative psychology, especially with respect to the most accomplished tool user, the chimpanzee. Possible directions were indicated by McGrew (1993), who listed twenty propositions concerning the intelligent use of tools. One proposition, for example, states that tool use, as it is observed in most nonhuman species, is highly context-specific, in the sense that it constitutes a habitual pattern displayed by those species. This is true for the digger wasp's pounding of the soil to plug nest-hole entrances or the sea otter's way of smashing mollusks. By contrast, chimpanzees, according to McGrew, "are the only non-human species in nature to use different tools to solve different problems. They go beyond using the same tool to solve different problems (e.g., a sponge of leaves to swab out a fruit-husk or a cranial cavity) or different tools to solve the same problem (e.g., probes of bark or grass or vine to fish for termites). Thus, they have a tool-kit" (McGrew, 1993, p. 168).

Another of McGrew's propositions is worth mentioning here, especially with regard to the concepts and evaluation of cognitive abilities advanced in the previous chapter. It appears that the degree of cognitive ability is not directly related to the complexity of tool use. For example, marine mammals such as dolphins do not use tools (Schusterman, Thomas, and Wood, 1986), although they do show elaborate communicatory and cognitive skills (see Chapter 6). Among nonhuman primates, different measures of cognitive abilities, provided for example by Piagetian tasks (see Chapter 3), usually cannot help to differentiate between the tool-using chimpanzees and the non-tool-using gorillas. McGrew notes that the only distinguishing intellectual correlate of tool use available so far concerns the ability of the chimpanzee, but apparently not of the gorilla, to recognize its reflection on the mirror. The exact competence that this ability connotes is a somewhat controversial topic, however, and it will be examined in Chapter 7.

If tool use has been observed in many species of animals, ranging from insects to birds and primates, it is worth asking how these tool-using activities might differ from a typically human use of tools. Although some animal species manipulate objects and prepare their tools, they apparently do not attempt to improve them in order to produce a definitive tool, as humans do. Furthermore, tool use is predominantly a solitary affair for animals (but see Boesch, 1993a), whereas human tool making and tool using are immersed in a network of social relationships and traditions. Thus, the human tool user not only manipulates tool objects, he also compares them so he can keep the "best" ones in a kind of "tool box." Moreover, humans do not necessarily abandon tools after use but may instead pass them on to others (for example, offer them as gifts) or exchange them with other individuals or groups.

5

Social Cognition

Cognitive processes in animals have been considered so far in the context of an individual responding to stimuli or objects in the physical environment. The examples described in the previous chapters have shown that external reality is filtered, analyzed, and processed by the organism. It is both parsimonious and convincing to deal with these phenomena at the behavioral level by inferring that representational processes underlie the responses. Complementary to the study of how animals act on their physical environment is the important question of how they manage their social lives and what types of cognitive organization sustain social behaviors. Because animals of most species are engaged in behaviors involving networks of social relationships, one can ask whether the individual's reactions to conspecifics recruit cognitive processes similar to those implied by the individual's treatment of inanimate objects.

According to several scholars (Jolly, 1966; Humphrey, 1976; Kummer, 1982), pressures for the emergence of cognitive processes in animals, particularly primates, are the strongest in the social domain, and thus intelligence may have evolved from needs originally arising from living in social groups (to make alliances, to compete, and so on). These needs imply, to various degrees, the "manipulation" of conspecifics. What relations can be established, then, between cognitive capacities used in the manipulation of the objects belonging to the physical environment and those involved in social interactions?

There is no compelling reason to postulate that the two kinds of capacities have to be radically different. One can in fact agree with Bates (1979) regarding the communicative and sensorimotor skills of human infants, that both social and nonsocial processes rely on a common "cognitive software." It would be too simplistic, however, to posit that the two domains and their respective kinds of "objects" do not differ. There are obvious differences between an inanimate object and a social (or animate) "object": when a subject acts upon an inanimate object, the latter *resists*, whereas an action on a social object produces, in most cases, a *reaction* or a response. These different cause-effect relationships imply that the instruments of action and the underlying regulatory mechanisms are different. The action of effectors (the hand or the foot, for example), guided by information from sensory organs, may be sufficient for handling an inanimate object; by contrast, a social object requires that the actor use additional "instruments," such as expressions, gestures, or vocalizations.

The idea that cognition is specific to a given context before it is generalized to other contexts has been considered by some theoreticians (Vygotsky, 1962; Fodor, 1975; Rozin, 1976), but this question cannot be addressed in the same way for humans and animals. A human's perception of the physical world is largely symbolic, in the sense that the world of objects is endowed with meanings and channeled through social rules. The symbolic content of perceptions is less evident when it comes to animals. According to some researchers (e.g., Cheney and Seyfarth, 1985, 1990a, 1992), animal intelligence reaches its full expression in the social sphere. Examples presented in the preceding chapters show, however, that animals demonstrate complex cognitive processing in their perception of *inanimate* objects. The problem of social cognition is made even more difficult by a lack of systematic studies of the idea that intelligence was selected primarily by the demands of a complex social life (Kummer, Dasser, and Hoyningen-Huene, 1990).

Numerous anecdotes are available to illustrate the social complexity of animal life. Byrne and Whiten (1988) have compiled an extensive catalogue of tactical deception in nonhuman primates. As an example, consider a case of deception in chimpanzees: these great apes often inhibit their intention movements and visual attention toward a desirable object in the presence of animals of higher rank; by "ignoring" the object, they may prevent others from finding it and therefore have access to it without competition later. In contrast to the availability of anecdotal reports, very few experimental studies have been done on this topic of social manipulation, which is central to cognition, partly because it is so difficult to investigate social objects and their relations.

Experimental Methods for the Study of Social Cognition

The experimental procedure of isolating a stimulus in order to vary certain aspects of it in a controlled fashion is commonly used for studying responses to inanimate objects (see Chapter 2 for examples). This procedure is of little use for studying social variables, however, because social variables are not easily manipulated. Two major problems arise. First, it is often very difficult to decide which is the most relevant modality among the plurality of stimuli intervening during a social exchange. Second, assuming that the most significant parameters are identified, it is usually impossible to extract those variables from the ongoing social context and interactions in order to evaluate their role experimentally.

Nevertheless, a number of experimental investigations, undertaken in laboratory conditions or in the field, have succeeded in isolating and manipulating some of these social variables. Although most of these studies have been carried out on mammals, and more particularly on nonhuman primates, there are some exceptions. Numerous experimental studies on social responses—individual recognition, prior evaluation of competitors, influence of social context, etc.—have been conducted in various species of fishes (Bronstein, 1983; Keeley and Grant, 1993). An abundant literature is available related to the attempts to teach mammals (predominantly chimpanzees) rudimentary forms of human language, by use of different media (gestural communication, arbitrary symbols). These studies will not be reported here (but see Chapter 6) for two reasons. First, they concern experiments in which the main goal is not to understand the communicatory context of exchanges between partners. In effect, even though these experiments require communication between members of different species (for example, between a human and a chimpanzee or a dolphin), this form of communication is very different from intra-species communicatory modes. Second, these studies have a limited impact in terms of social cognition because, as noted by Kummer et al. (1990), "the signs taught to chimpanzees contain no signs for kin, male, mother, or appeasement" (p. 85)—important "concepts" in animals' social groups.

In this chapter I describe studies that have tried to prove the existence—and possibly describe the forms—of social representation in nonhuman primates during natural exchanges with their conspecifics. The questions addressed will thus concern, in turn, the perception primates have of their social structures and whether they form "concepts" about interindividual relationships.

Social Cognition in Monkeys

Vocal Recognition among Vervet Monkeys

Much communication in nonhuman primates relays acoustic information. For this reason, interindividual exchanges of vocalizations are particularly useful for studying social representation.

Vocal recognition has been studied with free-ranging vervet monkeys in Amboseli National Park, Kenya (Cheney and Seyfarth, 1980). This study focuses on maternal recognition of offspring vocalizations; it illustrates the twofold advantages of testing animals under natural conditions and, at the same time, of employing criteria usually used in the laboratory (such as control conditions). The study attempted to address three questions: (1) Do female vervet monkeys react to the calls of their offspring? (2) Do mothers recognize their offspring on an individual basis? and (3) Can a female vervet recognize the offspring of other females?

Two groups were tested (at least seven juveniles in each age group between 1 and 3 years) with a playback method using intragoup vocalizations. Some typical calls of the young vervets, lasting approximately 7 seconds, were first recorded. The experimental setup requires that the mother of one of the experimental juveniles be out of sight of her offspring and in close proximity to two other females also having offspring in the group. When these conditions are met, a loudspeaker concealed in a nearby bush diffuses calls of the offspring of the mother subject. The behaviors of the three mothers are filmed for 10 seconds before and 45 seconds after the end of each call sequence.

The analysis is mostly concerned with the orientation of the mother's face toward the speaker. The main results indicate that both the mother and the control females look at the speaker when the vocalizations are played. The mother, however, tends to be quicker to look toward the speaker and to look for longer than the controls. These results provide a positive answer to the first question: mother vervet monkeys are indeed able, on the basis of voice alone, to classify juveniles into at least two categories, own offspring and other's offspring.

To answer the other questions, the researchers compared the reactions of control females with their behavior before each trial. The control mothers significantly increase their attention to the mother of the vocalizing juvenile. This result can nevertheless be explained in two different ways: either control females look at the mother because the latter reacts strongly to the vocalizations (in other words, increases in the mother's attention entail increases in control females' attention), or the control females recognize the calls of individual juveniles and thus know that a particular juvenile is associated with a particular female. Analyses of those trials in which the mothers *did not* orient toward the speaker helped the investigators choose between the two interpretations. In such cases, control females still looked at the mother, even though the latter showed no visible interest in the vocalizations of her offspring. In other words, control females look at the mother without any prior cue from the mother itself.

To summarize, Cheney and Seyfarth (1980) conclude that female vervets are not only able to distinguish their own offspring from the offspring of others, they are also able to recognize the offspring of other females. A similar demonstration is provided in a study (see Chapter 6 for a description) of recruitment of agonistic aid in rhesus monkeys (Gouzoules, Gouzoules, and Marler, 1984).

Social "Concepts" in the Macaque

Can a monkey perceive the equivalence between a conspecific and a picture of a conspecific? As the preceding example suggests, auditory signals play an important role in one monkey's recognition of another monkey. But vision, too, is a significant medium for the regulation of social exchanges in primates. A series of studies with long-tailed macaques has examined individual knowledge related to group structure (Dasser, 1987a,b, 1988). The general aim here is to explore the classificatory system used by the monkeys about their social structure. Do they, for example, as we humans do, establish classes of relationships? A more specific goal in one of the experiments is to test if female macaques have a "concept" of the mother-offspring relationship.

A prerequisite for using macaques in experiments requiring flexible methods and stimuli was to ascertain that group members can recognize pictures (color slides) of other macaques in the group. The first experiment (Dasser, 1987a) was conducted with the members of a group of 40 macaques living in a large indoor/outdoor compound (1,000 m²). Two females (4 and 5 years old) and one male (5 years old) were tested. After considerable training, the subjects were able to be separated temporarily from the group and placed alone in a test room, where they viewed slides of other group members. Two classical discriminative learning techniques are used, a two-choice simultaneous discrimination technique and a matching-to-sample technique (see Chapter 2). In brief, two-choice simultaneous discrimination requires each monkey to be trained to respond to one of two simultaneously presented stimulus animals; each stimulus is of a single slide showing the full face of a monkey. The stimulus animals are either familiar (a group member) or unknown (an individual from another colony). The subject indicates its response by pressing a button beneath the selected slide. In a transfer phase, the slides used as alternatives depict novel views of the faces of the same stimulus animals used in training, as well as views of the whole animals.

The results show that the two subjects tested with this procedure can, after training on a few samples, identify in test trials novel views of the animals used in training. A similar competence for individual recognition has been noted in studies with birds. For example, Ryan and Lea (1990) have demonstrated that hens rapidly learn to discriminate between two sets of slides depicting individual hens photographed from different angles. The birds can then transfer their discrimination to new slides of the same hens used in training.

To return to the experiment with macaques, once individual recognition is established, there follows a test with the matching-to-sample procedure. In this experiment, the monkey is presented with three slides. The first slide is located in the center of the apparatus and it shows the sample (positive stimulus) for a given trial. In training, the two comparison stimuli represent either the sample or an unknown macaque. In one of the transfer tests, the subject has to match slides of non-overlapping body parts: for example, the sample is the lower half of A's body and one comparison stimulus shows the upper half of A's body while the alternative depicts the lower half of B's body. The subject is food-reinforced whenever it selects the comparison slide that matches the sample. The single subject tested was reliably correct in the different transfer situations, including the matching of faces and of non-overlapping body parts. The minimal conclusion of this first study is that the three monkeys tested correctly identify members of their colony presented as color slides. In addition, the finding that one subject shows an ability to match non-overlapping body parts suggests that it uses stored information about the real animals and thus that it perceives "the equivalence between an object and its picture" (Dasser, 1987a, p. 72).

A "concept" of affiliation in the macaque. The possibility that monkeys have a social concept of affiliation may be formulated in this way: Can a monkey distinguish a mother-offspring dyad from other pairs of group members? The two female long-tailed macaques used in the previous experiment were tested again (Dasser, 1988). One subject is studied with the two-choice simultaneous discrimination procedure, the other with the simultaneous matching-to-sample (SMTS) procedure. In the SMTS task, the sample is a slide of a mother, the positive alternative a slide of her offspring, and the negative alternative another individual of the same age and same sex as the offspring. For training, two mother-offspring dyads are used as stimuli: 5 errors are allowed within blocks of 20 trials. A transfer test uses novel stimuli, mother-offspring dyads never previously presented. The possibility that recognition occurs because of mere physical resemblance between relatives, rather than because of their social relationship, is also tested, by showing slides of offspring taken up to three years earlier.

Matching is correct in 100 percent of the simultaneous discrimination trials and 90 percent in the SMTS trials. Both subjects experienced difficulties, however, in generalizing their matching to pairs when the offspring were shown at an earlier phase of development. A question can be raised about the results of this experiment: Do the two subjects simply classify members belonging to the maternal lineage from non-relatives? In fact, this hypothesis can be rejected, since in 8 out of 9 test trials (for the SMTS task) the negative stimuli represented individuals belonging to the same matriline, and the subject in this experiment clearly demonstrated her ability to differentiate relations within matrilines. For Dasser (1988), the capacities expressed by the two female macaques in discriminating and matching mother-offspring pairs reflect the use by these monkeys of an abstract category that may be analogous to the human concept of "mother-child" affiliation.

Dominance relationships. Dominance relationships appear to be an important factor in the lives of macaque monkeys; they seem, in particular, to ensure the cohesion of the troop (Deag, 1977). An observer can infer dominance relationships by noting the frequency, the intensity, and the outcome of agonistic interactions. With the method and one of the subjects used in the previous experiments, Dasser (1987b) experimentally investigated whether a female long-tailed macaque would show a "concept" of dominance in agreement with the observer's classification of dominance, established from agonistic relationships existing in the social group.

The test employs a simultaneous two-choice discrimination, as in the preceding experiments, and in the training phase the subject is taught to choose the slide of the dominant member of a pair. In a transfer phase, novel pairs of stimuli are presented. Two slides are presented on each trial: they picture the face of one of three adult females (always the same in the training) with the following dominance relationship: A > B > C. On each trial, female B is paired either with the top-ranking female A or with the low-ranking female C. The selection of the dominant member by the experimental subject is food-reinforced. The test phase includes the presentation of novel pairs of males or females (pairs are grouped by sex). A total of 28 transfer trials was presented, with one novel pair used on each trial. This procedure ensures that, because of different combinations of stimulus slides, the hierarchical position of 10 out of the 16 individuals shown is changed at least once during the experiment, thus precluding the possibility that the subject would learn always to choose a particular individual no matter which other individual is paired with it. The choices exhibited by the subject are far from straightforward: the dominant member is selected in only 5 pairs, and the subordinate in the remaining 23 pairs. Even though the results differ from chance, they are difficult to interpret, especially since the subject chose subordinate individuals in spite of not being rewarded for choosing them. Nevertheless, the data show that the monkey responded on the basis of an asymmetric social representation that corresponded to the dominance relationships in the group.

This experiment, along with those described above, suggest that the monkey does not simply react to a particular social behavior within its group (for example, an agonistic reaction for dominance). Rather, it reacts to the relation of dominance itself, and this is a reaction that also has internal properties, such as the transitive inference (see Chapter 3 for other examples of the mastery of this relation in nonsocial contexts by different species).

It must be kept in mind, however, that the data concerning social cognition are still coarse and that further experiments are required in order to draw more firm conclusions. But the available evidence gathered from nonhuman primates indicates not only that these animals know the individuals forming the groups in which they live but that they have a more abstract knowledge of these groups. This relatively abstract knowledge is evident in the social relationships of affiliation and dominance. It remains to be shown, however, what degree of abstraction is attained by these primates. In particular, one might ask whether, as it has been suggested (e.g., Thompson, 1995), some of the behaviors elicited by the test procedures could be instances of associative conditioning instead of concept formation.

Social versus Nonsocial Cognition

Primates interacting with objects in the laboratory face problems that, from a logical point of view, are identical to those that they confront in more natural situations. Even so, considerable differences are observed in their performance in the two contexts: for instance, knowledge related to the relative position of individuals in the hierarchy of the social group appears almost spontaneously in natural settings, while long training periods are usually required for this same kind of knowledge to emerge in laboratory conditions with inanimate objects (see Chapter 2).

As stated at the beginning of this chapter, one possible explanation for this difference in performance—the difference between processing information concerning inanimate objects and information relating to social objects—is that selection for intelligence has acted most effectively in the social domain. Thus, "during primate evolution, group life exerted strong selective pressure on the ability to form complex associations, reason by analogy, make transitive inferences, and predict the behavior of fellow group members" (Cheney, Seyfarth, and Smuts, 1986, p. 234). This idea is obviously seductive, but as noted by Kummer et al. (1990), the social-intelligence hypothesis is still in the stage of infancy. In effect, with the exception of the above data and a few other studies (see Cheney and Seyfarth, 1990a, for references on primates, and Zayan, 1994, for evidence of a perceptual transitive inference in the formation of linear hierarchies in birds), there is a clear lack of empirical evidence to support it.

In order to further explore the effect of selection pressures on social cognitive mechanisms, an investigator might systematically present problems of identical difficulty simultaneously in the social sphere (with conspecifics) and in the physical domain (with inanimate objects). This approach has not yet been undertaken, probably because of the undeniable difficulty of equating the complexity of social and nonsocial tasks. A slightly different way of examining this question would be to test monkeys or other animals with logically similar problems, using stimuli either social (conspecific) or nonsocial (animals of other species). As Cheney and Seyfarth (1985) put it, this question amounts to asking if vervet monkeys are as good naturalists as they are primatologists.

The answer to this latter question seems to be no. For instance, vervet monkeys may have good knowledge of the spatial distribution of their group members, but they apparently do not register the ranging patterns of other species. For example, vervets give acoustically distinct alarm calls in response to three kinds of predators, and playback experiments have demonstrated that these calls elicit different escape responses in the listeners (Seyfarth, Cheney, and Marler, 1980; the nature of these alarm calls will be further considered in the next chapter). In other experiments, vervets are presented with the calls of two species: the hippopotamus and a bird, the black-wing stilt, which lives in or near water. These calls, played from either a swamp or from a dry woodland habitat, do not affect the monkeys, which behave as though they do not recognize that the calls are not appropriate to a dry habitat. To sum up, "monkeys are primatologists who have spent too much time studying a single species, or living in the same group" (Cheney and Seyfarth, 1990b, p. 190).

Can we conclude from these field data that monkeys are systematically less able to acquire nonsocial knowledge than social knowledge? A problem of terminology arises here: can we equate, as Cheney and Seyfarth (1985) do, nonsocial stimuli such as a hippopotamus with an inanimate object? It is more likely that the hippopotamus is not exactly similar to an inanimate object, and it might be useful, in future discussions of social cognition, to consider at least three categories of objects: (1) social objects in the form of conspecifics; (2) other animate objects (animal of other species); and (3) inanimate objects.

The Use of Social Tools

The use of an inanimate object as a tool is certainly more rare than the use of other individuals as "social tools" to reach goals. A well-known instance of social tool use has been reported by Kummer (1971; see also Chapter 4): a female hamadryas baboon's use of her male leader as a tool in the so-called protected threat. The female, by presenting her rear to the male while simultaneously threatening another female, sometimes succeeds in inducing the male to attack her female opponent. The use of conspecifics as tools is apparently not limited to vertebrates. Kummer (1982) describes the case of weaver ants (reported by Lindauer) "which glue the edges of leaves together by holding a larva in their mandibles and by moving it back and forth between the leaves. By slight pressure, the worker ant causes the larva to secrete a silk thread which fastens the leaves together" (Kummer, 1982, p. 126). This example suggests that using conspecifics as tools phylogenetically predates the use of inanimate tools.

A colorful example of the use of a social tool comes from the daily life of the colony of chimpanzees living in the Arnhem zoo (de Waal, 1982, p. 47):

> On a hot day, two mothers, Jimmie and Tepel, are sitting in the shadow of an oak tree while their two children play in the sand at their feet . . . Between the two mothers the oldest female, Mama, lies asleep. Suddenly the children start screaming, hitting and pulling each other's hair. Jimmie admonishes them with a soft, threatening grunt and Tepel anxiously shifts her position. The children go on quarreling and eventually Tepel wakes Mama by poking her in the ribs several times. As Mama gets up Tepel points to the two quarreling children. As soon as Mama takes one threatening step forward, waves her arm in the air and barks loudly the children stop quarreling. Mama then lies down again and continues her siesta.

The observer of this social scene proposes the following interpretation, which takes into account two facts: (1) Mama is a high-ranking female; (2) conflicts between children regularly engender tensions between their mothers, probably because each mother generally attempts to prevent the other from intervening in a quarrel. This situation has most likely occurred here, and Tepel has solved the problem by activating a third party, Mama. This old and respected female rapidly understood her role as arbitrator and acted to prevent a potentially agonistic situation.

A remark must be made here about the concept of "social tool" as applied to the last examples. First, the baboon's "protected threat" is not tool use according to Beck's definition (see the previous chapter). In other words, the term *social tool use* may be used too generously; it is used in a metaphorical rather than a descriptive sense. Second, there is a distinction to be made between using a conspecific for a social goal (as the baboon did) and using a conspecific as an instrument for solving a physical problem (as weaver ants do). It is not obvious that using larvae in the way described should really count as social tool use, since the "tool" (larva) is so different from the users (fully developed ants). A recent example of the use of a social tool is now available: an adult female Japanese macaque was observed systematically sending her infant into a tube to recover food hidden in it (Tokida, Tanaka, Takefushi, and Hagiwara, 1994).

The Exchange of "Social Goods"

While humans frequently exchange altruistic acts for material goods (for example, giving a gift to acknowledge a favor), a similar flexibility of exchange patterns is not so obvious in nonhumans. For Seyfarth and Cheney (1984), nonhuman primates trade only in one "currency": one social act (e.g., grooming) for another (e.g., making an alliance). Such exchanges are frequently observed within matrilines, but they may also appear between non-relatives. For example, Cheney and Seyfarth (1990a) report that adult vervet monkeys are significantly more likely to react to the vocalizations of individuals with whom they have recently had altruistic exchanges (such as grooming) than to vocalizations of these same individuals if they have not recently had exchanges with them (see also Hemelrijk, 1994, for an example with long-tailed macaques).

A review of exchange modes in human and nonhuman primate societies indicates that exchanges among humans are, among other things, characterized by the introduction of objects (Bard and Vauclair, 1984). Using objects for social exchange is much less common among apes and other nonhuman primates. I will consider later (Chapter 6) how the social context and, more precisely, early communicatory exchanges between mother and infant contribute to the cognitive organization responsible for this use of objects (this kind of cognition will serve as further developmental support for communication and action on other objects).

Suggestions for Future Research

The systematic study of social cognition in animals has only recently begun; the emphasis has been put mainly on nonhuman primates, although more and more data are available for other taxa, such as birds (Zayan, 1994). The neglect of social cognition is partly due to an anthropomorphic bias in favor of animals' ability to manipulate inanimate objects (after all, most human psychological studies focus on reactions to nonsocial stimuli). One should not forget, however, that methodological difficulties complicate the study of social cognition.

In fact, the available record on questions of social cognition can help guide the future research program of the field. This program should address three major points (Cheney et al., 1986). First, there is an urgent need for data on more primate and nonprimate species. Second, there is also a great need for in-depth investigations of the relation, for a given species, between cognitive capacities in the laboratory and those expressed in naturalistic social settings. Results from such experiments could lead, in the short term, to predictions of qualitative differences in social relationships across species. Third, the study of spontaneous social relationships should be supplemented by investigations of animals' knowledge in the physical (i.e., nonsocial) domain. Cognitive mapping (see Chapter 4) and food-searching behaviors are good candidates for such studies, because of their immediate ecological validity for the animals.

The three axes just mentioned are indispensable for increasing our understanding of social cognition, as well as the cognitive processes expressed in other domains. "This is essential if we are to compare social intelligence with intelligence in other domains, and if we are to test the intriguing hypothesis that primate intelligence—including our own—originally evolved to solve the challenges of interacting with one another" (Cheney et al., 1986, p. 1365).

Summary and Current Debate

The early sections of this chapter stressed the difficulty of studying social cognition. Although considerable anecdotal evidence supports the contention that cognition is at its most developed in the social domain (especially among nonhuman primates), experimental data are needed to strengthen this claim. We must keep in mind that the most impressive data so far obtained on social cognition in primates (by, for example, Cheney and Seyfarth and Dasser) were gathered from small numbers of subjects, are barely significant from a statistical point of view, and thus have yet to be replicated.

Quick mention might be made of an alternative hypothesis for the evolution of advanced cognitive abilities in nonhuman primates—namely, that increased foraging demands led to elaborated cognitive abilities, which in turn (or in parallel) allowed greater possibilities for social interaction.

For example, cranial capacities (and the related capacity for learning) in primate species are more likely to have evolved in frugivores or omnivores than in folivores (Milton, 1981). Similarly, species engaged in extractive foraging of embedded food resources may have developed a brain larger than that of non-extractive foragers (Gibson, 1986).

One current debate between ethologists and psychologists in fact concerns the use of descriptions (favored by ethologists) and how to integrate them with the experimental data favored by comparative psychologists. A case in point is primate cognition. I described earlier in the chapter how Kummer et al. (1990) have urged that we should systematically rely upon experimental procedures in the study of social cognition. This position is at odds with the tradition established in ethological studies of mixing observational data with experimental results. In fact, the problem reiterates the traditional dichotomy between field and laboratory—in other words, the debate over methodology between the naturalist and the experimentalist. Writing in response to the article by Kummer et al. (1990), de Waal (1991) proposes that naturalistic approaches might be reconciliated with experimental investigations through a complementary methodology in the study of social cognition. This methodology includes qualitative descriptions, such as anecdotal reports (e.g., on deception), quantitative descriptions, controlled observations, and experimentation.

The rationale for this opportunistic approach rests on several arguments; only some of them are listed here. For example, research on social cognition needs qualitative descriptions because cases of social intelligence often manifest themselves in successful solutions to unusual problems. In other words, some behaviors will not so readily be elicited by quantitative methods or experimentation. Moreover, accounts of rare social events can serve to dramatize questions about behavior and to uncover variables that could later be manipulated in experiments. If the burden of proof for the existence of a cognitive capacity ultimately rests on the control of variables, it remains the case that some social phenomena cannot transfer to the laboratory. For example, the context in which the use of "social tools" by the chimpanzees of the Arnhem zoo (see above) has been studied would probably be impossible to re-create in the laboratory. As a consequence of this limitation, "students of social cognition need to develop a methodology that negotiates between the richness of context and the multitude of social options open to the animal under natural conditions and the more controlled but also limited experimental setting" (de Waal, 1991, p. 315).

In sum, the evidence gathered so far has shown that monkeys are able to classify relationships and even compare relationships involving different individuals. If animals appear to construct and use social representations about the behavior of others and to anticipate the consequences of their acts, may we conclude that they attribute motives or intentions to others? This question asks whether they have attributional capacities, or whether they have a "theory of mind" (Premack and Woodruff, 1978).

The study of the attribution of mental states in animals has been essentially restricted to nonhuman primates, especially chimpanzees. Some of these studies will be presented and discussed in the following chapter because of their obvious link to language. Examination of the attributional capacities themselves will be deferred until Chapter 7.

6

Animal Communication
and Human Language

What is the status of animal communication with respect to human communication and, more specifically, with respect to human language? Before we attempt to compare animal and human communication skills, we must start with a concrete understanding of the features of animal communication and a critical examination of the performance of language-trained animals (marine mammals and primates). The first step is to define clearly both the key concepts we use to describe communication and the contexts in which these concepts—communication, representation, and language—intervene.

Representation is defined as the phenomenon by which an organism structures its knowledge with regard to its environment. In the context of linguistic studies, and since de Saussure (see below), knowledge implies a support or a substitute in the knower's mind for the object in the external world. Two types of substitute can be distinguished: internal substitutes (such as indices or images) and external substitutes (such as symbols, signals, or words).

Communication consists of exchanges of information between a sender and a receiver using a code of specific signals that usually serve to meet common challenges (reproduction, feeding, protection) and, in group-living species, to promote cohesiveness of the group.

Language is a system that is both communicatory and representational (Vygotsky, 1962; Bronckart, in Zivin, 1979). It is grounded in a social convention that attributes to certain substitutes (the words) the power to designate the content (objects) of the communication.

For the purposes of this discussion, these definitions must be qualified. First, representation is understood here as a specific cognitive function that relates two entities: an object (concrete or abstract) with its substitute, or representation. Second, my definition of communication corresponds to that generally used by ethologists (e.g., Manning and Dawkins, 1992). And third, representation and communication will be considered separately for most of this chapter. In effect, one can conceive of communicatory events that do not possess representational elements. Thus, some nonverbal communicatory sequences in humans, such as the vocalizations emitted by a baby to signal a state of discomfort to its mother, certainly do not involve representation. In most cases, however, representational elements are a necessary component of the phenomenon of communication, at least in humans.

The previous chapters have stressed the fact that several animal species use different kinds of representations and supported the contention that "mental organizations exist in animals, in particular in nonhuman primates, as a necessary part of the perception of objects and their localization and interrelationships in space and time" (Walker, 1983, p. 380). Such a perspective makes reasonable the hypothesis of evolutionary continuity in the processing of cognitive information. If continuity in the processing of perceptual information, of space, and of memory indeed exists, it is tempting to assume that a similar continuity may exist with respect to the kinds of processing required for the emergence of language. This assumption depends of course on the definition one chooses for language. Piaget's conception of language, for example, is in obvious agreement with the theory of continuity (on Piaget, see chapter 3). For him, language is not a specific entity but a particular aspect of the general capacity for representation: "Language is only one aspect of the symbolic (or semiotic) function. This function is the ability to represent something by a sign or symbol or another object. In addition to language the semiotic function includes gestures . . . deferred imitation . . . drawing, painting, modeling" (Piaget, 1970, p. 45).

Other theories of the origins of language, in addition to Piaget's, also may shed light on possible continuities between animal and human cognition. For example, for Hewes (1973), human language derives from a system of vocalizations or gestures analogous to the systems of communication observed in some present-day animal species. And for Lieberman (1984), the organization of language (for example, its syntax) is seen as a generalization of the mechanisms that control nonverbal behavior.

Comparisons of Animal and Human Communication

One way of comparing communication abilities of different species with our own is to confront animals with ostensibly the most sophisticated form of human communication—namely, language. This approach entails listing the basic traits of human language and then searching for possible equivalents in animal communicatory systems. Hockett (1960), for example, identified 13 design features of human language that may in theory be present in any other communicatory system (see Table 6.1).

According to Hockett, features 1 to 5 (vocal auditory channel, broadcast transmission and directional reception, rapid fading, interchangeability, and total feedback) are shared by many animal species, among them several birds and most of the mammals; specialization, semanticity, and

Table 6.1. The thirteen design features of language (from Hockett, 1960).

1. Vocal auditory channel

2. Broadcast transmission and directional reception

3. Rapid fading (transitoriness)

4. Interchangeability: a speaker can reproduce any linguistic message he can understand

5. Total feedback: the speaker of a language hears everything of linguistic relevance in what he himself says

6. Specialization: sound waves of speech serve only as signals

7. Semanticity

8. Arbitrariness

9. Discreteness

10. Displacement

11. Productivity

12. Traditional transmission

13. Duality of patterning

arbitrariness (features 6–8) can be found in primates only, whereas discreteness and traditional transmission are specific to hominoids. For Hockett, the features of displacement, productivity, and duality of patterning appeared with human speech only. These three features are important and deserve some explanation. *Displacement* describes the ability to evoke things that are remote in space and time; *productivity* concerns the ability to "say things that have never been said before and yet to be understood by other speakers of the language" (Hockett, 1960, p. 90). *Duality of patterning* relates to the two-level organization of language: words as meaningful units or "morphemes" (level 2) are made of meaningless sounds or "phonemes" (level 1). These level-1 units are very limited in number and their meaning emerges from their permutations. For example, *tack, cat, act* are totally distinct as to meaning, and yet they are composed of the same sounds. Hockett's list was extended by Thorpe (1972), who added some features, such as prevarication and deceit, reflexivity, and metalanguage.

Of course, with this approach one can be easily biased by the traits one chooses to emphasize. It is nevertheless worth mentioning these features because of their direct relevance to the theoretical question posed by the comparison of animal communication and human language. Arbitrariness and deceit will be examined later, in the sections concerned, respectively, with language-trained apes and the question of intentionality in animals.

Signaling Behaviors in Insects

Displacement, as I mentioned before, concerns the property of language that makes it possible for the speaker to refer to objects or events that are remote in time and place and also to talk about things that have no spatial localization or that can never occur. In other words, a crucial property of the linguistic sign is that it be detached from the element (object, event, or state) to which it relates, and that its meaning be available regardless of the contextual situation.

Von Glaserfeld (1977) argues that displacement alone fails to achieve this transformation, because a mere delay (distance in time and space) does not change the one-to-one correspondence between the sign and the situation. In brief, a linguistic entity not only relates an object to a sign but also relates signs to other signs. The famous dance of honey bees, which conveys information about the direction and the distance of a food source (Von Frisch, 1950), provides a good illustration of the requirements for the process of symbolization. Only a hypothetical bee "observed to communicate about distances, directions, food sources, etc., without actually coming from, or going to, a specific location" (Von Glaserfeld, 1976, p. 222) could be said to use symbols. Similarly, while bees convey information about the environment, they do not generalize the use of their system to say anything else with it.

Signaling Behaviors in Birds and Mammals

Chickens emit vocalizations when they encounter food, and males are especially vocal when they encounter females. On finding food items, male domestic chickens produce "food calls" that vary with the type of food: the number of food calls varies with the preference ranking of the food. Moreover, a food-calling male in the presence of a female chicken is more likely than a silent male to elicit the female to approach (Marler, Dufty, and Pickert, 1986a). This research indicates that there is a strong association between the perception of the food by the signaler and the production of signals.

One can ask whether signals related to food only have a purely informative function, as it is the case with bees' dances, or whether they might take on a more representative or semantic function. Konishi (1963) showed that calls identical to food calls are emitted by birds when individuals are temporarily separated from their familiar companions and seek to reestablish contact with them. In other words, these signals can be used, to a relatively limited extent, in a context slightly different from the one in which they are usually emitted. As will be shown later (see the section on the manipulation of information during communicative exchanges in the next chapter), the signals can be modulated according to the identity of the receiver and can even be emitted to deceive conspecifics.

Birds also produce different alarm calls in response to terrestrial and aerial predators (Konishi, 1963). Playback experiments were conducted with bantam chickens (Evans, Evans, and Marler, 1993). Given that these birds are responsive to video images, they are shown tapes depicting an aerial predator (a hawk) and a ground predator (a raccoon). The visual images are sufficient to elicit alarm calls in the tested birds. Subsequent playback of aerial and terrestrial alarm calls produces qualitatively different responses. The former calls cause the birds to run toward cover and look up, while the latter cause them to adopt a highly erect "vigilant" posture. These findings suggest that the signals are in part functionally referential, since adapted responses can be elicited in the absence of contextual information (for example, the nonvocal behavior of the sender).

Field observations of ground squirrels (Owings and Virginia, 1978) reveal that these mammals give acoustically distinct calls to various classes of predators (for example, to hawks). Moreover, field playback of these calls elicits the same escape responses as those observed when the predator is actually present (Leger, Owings, and Boal, 1979).

It also appears that nonhuman primates can use communicatory signals whose meaning seems to be relatively detached from their context of production. A well-established example concerns alarm calls produced by vervet monkeys upon detecting predators. Vervet monkeys in Amboseli National Park (Kenya) have three main classes of predators (leopards, pythons, and eagles) whose presence is signaled by different alarm calls (Strushaker, 1967; see Figure 6.1). The emission of each type of alarm call evokes a different and appropriate response in conspecifics. For example, when monkeys are on the ground, eagle alarms cause them to look up and run into dense bush, apparently to avoid the eagle's swoop. A similar classification of alarm calls as a function of the class of predator has been described for lemurs, a type of prosimian primate (Pereira and Macedonia, 1991).

In field experiments using playback techniques (see Chapter 5), recordings of alarm calls have been broadcast to groups of vervets in the absence of predators (Seyfarth et al., 1980). It was found that alarm calls, broadcast through a loudspeaker, produce the same kind of adapted responses as those elicited by the presence of a predator. Furthermore, the responses observed appear to be independent of variations in call length and amplitude, as well as of the arousal state and age of the signaler. This experiment—which indicates that these signals not only indicate the sender's emotional state but also refer to external objects or events—speaks in favor of the interpretation of alarm calls as rudimentary semantic signals. Investigations of other types of calls in wild vervets, namely social grunts (such as grunts emitted when approaching dominant or subordinate individuals), reveal that these calls function to designate external objects (Cheney and Seyfarth, 1982).

Similar findings come from experiments with macaques. In one experiment (Gouzoules et al., 1984), screams of immature rhesus monkeys are tape-recorded to be played back to their mothers later from a hidden speaker. These calls function as "designators" for different categories of

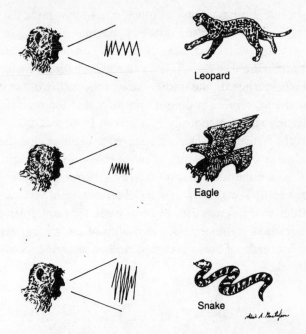

Figure 6.1. Alarms calls in vervet monkeys. Vervets emit different alarm calls, indicated here by different sound spectrograms, in response to the sight of different predators. Other vervets respond accordingly to the alarm call: in response to leopard alarm calls, they tend to run up trees; in response to eagle alarms, they look up or seek cover; in response to snake alarms, vervets on the ground are likely to stand bipedally and search the ground around them. (From Quiatt and Reynolds, 1993; reprinted with the permission of Cambridge University Press.)

opponents. For example, a "noisy scream" is usually emitted when a monkey encounters higher-ranking individuals and is accompanied by physical contact, while a "pulsed scream" is directed to relatives. Four scream classes were played to the mothers, which responded by looking to the speaker. Differences in response durations were observed as a function of the type of call. "Since screams were played to subjects in the absence of any other features of the original agonistic event, notably the caller and its opponent, the information necessary for differential responses is apparently contained solely in the vocalizations themselves. The scream vocalizations appear to have, as referents, the type of opponent and the severity of aggression in agonistic encounters: these are external referents" (Gouzoules et al., 1984, p. 190). Moreover, these screams may recruit assistance during certain confrontations, by informing the signaler's potential allies about the nature of the opponent.

Although the playback technique is an astute one for experimenting in natural or seminatural settings, it has a logical limitation: "By focusing on the behavior of recipients, the studies show us the nature of the inferences formed by the recipients but do not inform us that the communicator is communicating representationally" (Snowdon, 1990, p. 232).

Some of the features that Hockett reserved as defining characteristics of *human* language (see above) are somewhat challenged by recent findings on natural communication in animals. In particular, it appears that the crucial feature of semanticity can be observed, with various levels of sophistication, in the signals used by some birds and nonhuman primates. The other features (arbitrariness and duality of patterning) retained by Hockett as hallmarks of human language will be discussed toward the end of this chapter.

Language-Trained Animals

According to many prominent linguists, such as Chomsky (1968), human language is a unique phenomenon, made possible by the language "organ" of the brain, without analogue in the animal kingdom. This opinion is shared by a number of psychologists and anthropologists. Notwithstanding this claim that language is a distinctively human attribute, a number of investigators have attempted to explore the capacities of animals (mostly nonhuman primates) to learn to use at least some elements (syntactical rules, word lists) of human language. Some of these studies are summarized below.

One of the first attempts to teach apes the rudiments of language involved a home-raised chimpanzee named Vicki, who was taught to utter English words. After years of training, Vicki was able to produce four words with difficulty, *papa, mama, cup,* and *up* (Hayes, 1951). "Speaking" apes have such limited success because their phonatory apparatus is ill adapted to the articulation of the sounds of human speech (Lieberman, 1984). The main research projects on language with apes are listed in Table 6.2.

Teaching Gestures to Apes

One way to circumvent the articulatory limitations of nonhuman pimates is to use the natural predisposition of apes and especially of chimpanzees

Table 6.2. The main research projects on language in great apes.

Researcher	Species and name	Signals used
Hayes	Chimpanzee (Vicki)	Spoken English words
Gardner	Chimpanzee (Washoe)	Gestures
Terrace	Chimpanzee (Nim)	Gestures
Patterson	Gorilla (Koko)	Gestures
Miles	Orangutan (Chantek)	Gestures
Premack	Chimpanzee (Sarah)	Plastic chips
Rumbaugh & Savage-Rumbaugh	Chimpanzees (Lana, Austin, Sherman); bonobos (Kanzi, Mulika)	Lexigrams
Matsuzawa	Chimpanzee (Aï)	Lexigrams

to gesture. A systematic project to train a female chimpanzee named Washoe to communicate with ASL (American Sign Language) was thus undertaken (Gardner and Gardner, 1969; Gardner, Gardner, and Van Cantfort, 1989).

After three years of training, Washoe could use 68 gestures, mostly in the context of injunctions ("More!" "Come!" "Out!"). The size of Washoe's vocabulary was later extended to around 150 "words." These strings of gestures could be combined into sequences of three or four elements (*/you/me/go out/hurry/*) in highly restricted contexts, such as requests for food or for playing. An attempt to teach the same gestural language was undertaken with another chimpanzee, Nim (Terrace, 1979; Terrace, Petitto, Sanders, and Bever, 1979). Detailed analyzes of multisign utterances by Nim and Washoe were conducted to search for syntactical regularities. The researchers (Terrace et al., 1979) failed to find any evidence of an ability to use even an elementary form of grammar. Moreover, videotapes of the gestural signs produced by Nim and Washoe indicate that their gestures are expressed in two main contexts: after an injunction by the trainer or to request something. Additionally, many of the chimpanzees' gestures appear to be mere imitations of the teacher's prior utterances (Terrace, 1985, and see Figure 6.2). Terrace's arguments have been refuted forcefully (Gardner et al., 1989). It was argued, for example, that Nim was tested in highly restricted training situations that were not conducive to communication. Moreover, Washoe's spontaneous utterances in free play situ-

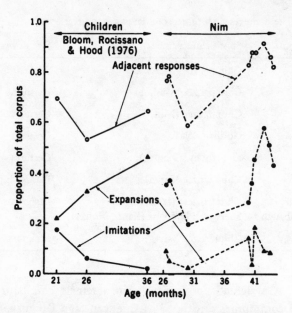

Figure 6.2. A comparison of types of utterances from children and from Nim (a chimpanzee). Proportion of utterances emitted by children *(left)* or by Nim *(right)* that are adjacent to, imitative of, or expansions of an adult's prior utterance. (*Note:* Adjacent utterances are those that follow, without a pause, an adult utterance.) (From Terrace et al., 1979.)

ations were much more frequent than Nim's. Finally, the contexts in which these gestures are produced by the apes also differ from the contexts in which humans produce their first utterances (see Figure 6.2).

Attempts have been made to investigate the gestural communicatory competence of other ape species: for example, the gorilla Koko (see Patterson, 1978, and Patterson and Linden, 1981) and the orangutan Chantek (see Miles, 1983, 1990).

Psycho-Linguistic Studies in Marine Mammals

Marine mammals (dolphins, sea lions, otters) are among the largest-brain mammals; their brains compare well with primates' enlarged brains (Jerison, 1973). It is therefore not surprising that these species are an interesting model for the investigation of linguistic-like capacities of species phylogenetically close to humans. Two projects, with sea lions and dolphins, are relevant in this regard. California sea lions are trained in a symbolic matching technique using gestural signs (Schusterman and

Krieger, 1984; Schusterman and Gisiner, 1988; Gisiner and Schusterman, 1992). These signs, for the most part arbitrary with respect to their assigned meaning, are produced by arm and hand movements of a trainer who is seated at the edge of a pool (see Figure 6.3). The signs refer to types of objects (pipes, balls, rings, frisbees, etc.), to modifiers (size, color, and locations of objects), and to actions (mouthing, placing, touching, etc.). Two sea lions are trained with three-sign constructions consisting of modifier + object + action. Upon presentation of the sequences of signs by the trainer, the subject leaves to execute the signed sequence. After 24

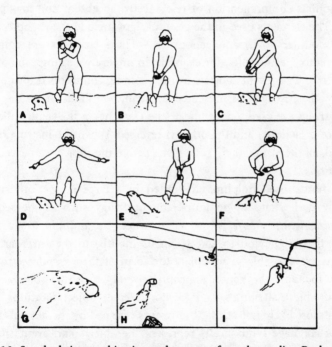

Figure 6.3. Standard sign combination and response from the sea lion Rocky. *A:* The object sign WATER is given. *B:* The signaler pauses (indication of the separation of two objects in the relational sign combination). *C:* The modifier WHITE is given. *D:* The modifier SMALL is given. *E:* The object sign BOTTLE is given (Rocky searches for the object designated by the combined modifier and object signs). *F:* The action sign FETCH is given (Rocky remains at station until released by the signaler's foot dropping). *G:* Rocky goes to the bottle and starts to move it while scanning the pool for the destination it must reach (a stream of water). *H:* Rocky approaches the stream of water while pushing the bottle. *I:* Rocky places the bottle in contact with the stream of water. (From Gisiner and Schusterman, 1992; copyright 1992 by the American Psychological Association. Reprinted with permission.)

months of training, one of the sea lions can comprehend 20 signs (5 modifiers, 10 objects, and 5 actions) in 190 three-sign combinations. For example, to the request /Black/ball/mouth/ the subject will go over to the black ball (and not to the grey or white balls) and place its open mouth on it.

Experiments with dolphins follow the same basic procedure, with the aim of studying the subjects' comprehension of imperative as well as of interrogative strings. These strings are made of arbitrary gestural signs or of computer-generated sounds (Herman, Richards, and Wolz, 1984; Herman, 1986; Herman et al., 1993). The interrogative form allows testing of the dolphin's comprehension of reference to an absent and to a present object. The elements used in the gestural signs include agents, objects *(O)*, object modifiers *(M)*, and actions *(A)*. These elements are combined according to specific rules: for example, in imperative strings, the order is $M + O + A$. To prevent subjects from becoming conditioned to the linear order of the instructions presented to them, two kinds of arrangements of the strings are used: one is linear (the elements in the string follow the sequence of actions) and the other is reversed. When the instructions are reversed, all the elements must be given before the subject can execute the instructions.

Two bottlenosed dolphins are tested in a large tank. Both subjects correctly understand up to four-element strings, such as the sequence /Surfboard/right/frisbee/fetch/, which instructs the dolphin to go to the frisbee to its right and to take it (in its mouth) to the surfboard. The dolphins are also able to modulate the form of their responses to given actions, to apply the action appropriately: (1) to novel objects, (2) to different object attributes, and (3) to different object locations. These abilities most likely reflect that the "words" learned by the animals in this artificial language symbolically represent the objects and events referred to in the strings of gestures.

Early works with the dolphins came under the critique of some psychologists, like Premack (1985), who questioned whether "words" and "sentences" that were given to the dolphins through motor commands qualified as the kind of code usually described as a "language."

Teaching an Artificial Language to Apes

Investigators have employed specially designed artificial media to study the linguistic capacities of apes. Three research groups—two American (see the following references to Premack and Rumbaugh and Savage-Rumbaugh) and one Japanese (see, for example, Matsuzawa, 1985, 1987)—have been responsible for much of the work in this area.

Project Sarah. In studies carried out by Premack (1971), a female chimpanzee named Sarah was trained to manipulate plastic chips (tokens) differing in color and size. The plastic chips are backed with metal so that they can adhere to a magnetic board. Each token stands for a specific object *(apple, pail),* for actions *(give, take, insert),* qualifiers *(red, yellow),* concepts and conditionals *(same, different, color of, if-then).* Sarah was systematically trained to place the tokens on the board in order to obtain physical and social rewards from the trainers. For example, in order to receive an apple, Sarah had to place the two tokens *give* and *apple* in a vertical sequence on the board. Sarah's understanding of more complex sequences, such as the conditional, was tested, and it was shown, for example, that she is able to learn and differentiate the string:

/If/Sarah/take/apple/then/Mary/give/Sarah/chocolate/

from the string:

/If/Sarah/take/banana/then/Mary/no/give/Sarah/chocolate/.

In one of the experiments, Sarah is tested for her capacities to analyze the features of an object. Sarah is first shown a real apple and is then given a series of paired comparisons that describe the features of the apple: for example, red versus green, round versus square. Sarah has to pick the descriptive features that belong to the apple. Next, Sarah is presented with the blue plastic triangle that stands for the object apple and is again given the paired-comparison test. Sarah correctly assigned the same features to the token "apple" that she earlier assigned to the real object. These results indicate that even though the apparent features of the symbol are contradictory with the features of the corresponding object (a blue object standing for a red object), the chimpanzee correctly describes the apple. Sarah's performance reveals "that it was not the physical properties of the word (blue and triangle) that she was describing but rather the object that was represented by the word" (Premack, 1972, p. 98). In similar situations, the chimpanzee was able not only to associate objects, actions, or attributes of objects to specific tokens but also to associate one token to another token.

Lana, Sherman, and Austin. A different artificial medium was used in a

study with Lana, another female chimpanzee. Lana is trained with geometric symbols that light up on a keyboard when touched (Rumbaugh,
1977). To obtain a reward, the subject touched these arbitrary visual
symbols, called lexigrams, in a predetermined order, for example, according to the sequence */please/machine/give/banana/*. Two additional chimpanzees (Sherman and Austin) were subsequently instructed with these
lexigrams. An example of the work carried out with these chimpanzees is
a study (Savage-Rumbaugh, Rumbaugh, Smith, and Lawson, 1980) that
suggests the existence of representational symbolic abilities in apes. The
results of this study parallel the data reported above regarding Sarah's
competence in using substitutes for objects and for object attributes.

In the *first phase* of the experiment (see Figure 6.4), the chimpanzees

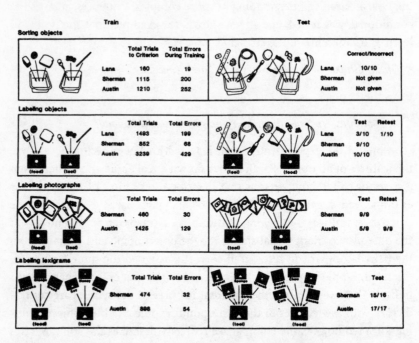

Figure 6.4. In an experiment on functional and symbolic categorization,
chimpanzees learned the items on the left and were tested, in a blind setting, with
the items on the right. The numbers of trials and total errors are given for the
training period, and the number of correct trial-1 selections is given for the testing
period. In the final labeling test, 17 different lexigrams controlled for physical
similarity to the categorical lexigrams were used. (Reprinted with permission from
Savage-Rumbaugh et al., 1980; copyright 1980 American Association for the
Advancement of Science.)

are trained to classify objects: three food items (e.g., a beancake, an orange) are sorted into one bin and three tool items (e.g., a stick, a key) are sorted into another bin. Then, after the subjects make correct reponses 90 percent of the time with novel food and tool items, they are presented with the lexigrams representing either "tool" or "food." In this *second phase* (labeling objects), the subjects are required to sort a tool or a food into the proper bin and then to select the lexigram representing either category. The presence of this labeling skill is tested by introducing additional foods and tools.

In a *third phase*, Sherman and Austin are requested to classify photographs of the original objects. As previously, a generalization test is proposed by introducing novel photographs. The final and crucial phase of the experiment consists of labeling lexigrams. First, the subjects classify the original lexigrams and, once the 90-percent criterion is reached, they are tested with lexigrams for novel objects. Sherman is correct for 15 items out of 16 and Austin is correct for 17 of 17. The authors conclude that Sherman and Austin are "able to treat 'food' and 'tool' as representational labels, and to expand the use of these labels to other exemplars because of training which encouraged the appearance of functional symbolic communication between chimpanzees" (Savage-Rumbaugh et al., 1980, p. 924). Additionally, this ability to manipulate symbols may be indicative of the chimpanzee's mastery of *reference*, a linguistic essential (for critical perspectives on the interpretation of the above data, see Epstein, 1982, and Wallman, 1992, pp. 71–74).

Whatever the true linguistic status of these chimpanzees' behavior, it is quite evident that the performances of Sarah and Sherman and Austin indicate the chimpanzees' ability to treat a representation (symbol) of an object as if it were the object.

Studies of bonobos. The referential abilities of apes have been further investigated through work with two pygmy chimpanzees (or bonobos), Kanzi and Mulika (Savage-Rumbaugh, Rumbaugh, McDonald, 1985; Savage-Rumbaugh, McDonald, Sevcik, Hopkins, and Rupert, 1986). Unlike Sherman and Austin, the two bonobos began to use lexigrams for the purpose of communicating with no specific training. For more than two years, Kanzi was playing in the room where Matata (his adoptive mother) was taught lexigrams (Rumbaugh, Savage-Rumbaugh, and Sevcik, 1994). When Kanzi was 30 months old he was left alone, and from the first day of Matata's absence, Kanzi was observed *spontaneously* (with neither specific food reinforcement nor specific language training) pressing lexigrams on the keyboard to request food, objects, and actions (e.g., tickling, chasing). Kanzi is thus able, by combining presses on lexigrams and gestures, to ask for desired objects or events and to tell the names of items in response to queries by the teacher. And contrary to the experience with Nim (see above), 80 percent of Kanzi's productions were spontaneous (see Figure 6.5). Nevertheless, the great majority of Kanzi's and Mulika's utterances expressed some sort of request "to direct teacher's attention to places, things and activities" (Savage-Rumbaugh et al., 1985, p. 658).

Additional analyses of Kanzi's 1,422 utterances combining two or more elements, recorded over a five-month period when Kanzi was five-and-a-half years old, suggest some form of protogrammar (Greenfield and Savage-Rumbaugh, 1990). Roughly speaking, half of these gesture-lexigram combinations qualify as spontaneous—that is, they are not responses to a human caregiver's utterances or simple imitations of gestures performed by others. Among the criteria used by human developmental psychologists to assess the existence of a protogrammar is the requirement that each

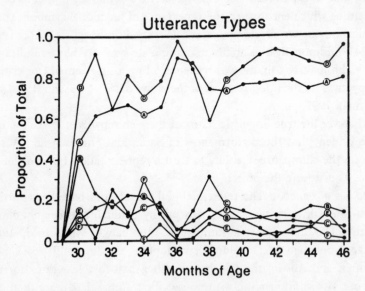

Figure 6.5. Types of utterances produced by the bonobo Kanzi. *A:* Single "words" used spontaneously. *B:* Single "words" elicited by companion's queries. *C:* Single "words" used in imitation or as result of prompting. *D:* Combinations used spontaneously. *E:* Combinations elicited by companion's queries. *F:* Combinations used in imitation or as a result of prompting. (From Savage-Rumbaugh et al., 1986.)

element in a combination must have an existence independent of the combination. Thus, when Kanzi requests a banana using the /*banana*/ lexigram, the chimpanzee is presented with a number of selected fruits and his choice must agree with the symbol that he has produced earlier (Kanzi is correct on 80 percent of such choices). A second criterion to assess the presence of grammar is that there must be rule-governed ordering of the elements expressing some stable meaning. Almost all of the bonobo's productions satisfy this rule (for example, Kanzi rarely repeats himself). Finally, Kanzi has learned the symbol order used by the caregiver—namely, the fact that the action precedes the object—in addition to the use of a personal rule. For example, to indicate the relation between agent and action Kanzi places the lexigram (e.g., /*bite*/) before the gesture (e.g., pointing toward a caregiver).

The authors conclude that Kanzi's achievements differ from children's grammatical development. First, his development is much slower: Kanzi needed three years to achieve the grammatical progress typically found after about one year in humans. Second, Kanzi is able to produce many fewer combinations than a child can after three years of speech. Third, contrary to human children, who use language to make indicative or declarative statements, 96 percent of Kanzi's productions are requests. This limitation in the chimpanzee's types of productions might be due in part to constraints inherent to the experimental environment. In effect, this environment strongly encourages Kanzi to make requests for activities or objects.

Differences in the Use of Signs by Apes and Children

The examples of the use of symbols by apes in the previous section make one wonder about the language abilities of the human child, especially the child in the process of acquiring language. What differences are there between a child's use of signs and the use of reference in expressions, made either spontaneously or after training, by linguistically trained apes?

The displacement feature, as described by Hockett (1960), is rarely completely fulfilled, if at all, in natural animal communicatory systems (see above). This limitation also apparently applies to language-trained chimpanzees, given that these animals generally use what they learn to attain immediate goals: the referents occur simultaneously with behaviors directed toward their objects. By contrast, human children's first utterances are characterized by both a disengagement from context and spontaneity. In other words, naming reflects an intent to communicate for its own sake (for example, to communicate the fact that the child has noticed an object or that objects have names) and is not solely the expression of a demand (Bates, 1979). Terrace (1985) provides a comprehensive account of the similarities and differences in how and in which contexts a child and an ape use symbols.

Distinguishing between Imperative and Declarative Functions of Language

In more general terms, a major difference between humans and apes is that the use of a "word" by apes is largely restricted to its imperative function, whereas humans will also use a word as a declarative. Declaratives (Bates, Camaioni, and Volterra, 1975) can be words or gestures, and they "function not primarily to obtain a result in the physical world, but to direct another individual's attention (their mental state) to an object or event, as an end in itself. Thus, a human toddler might say 'Plane!' apparently to mean, 'It's a plane!' or, 'Look! A plane', and so on. Here, the child communicates simply to share interest in something" (Baron-Cohen, 1992, p. 149).

It can be asserted with some confidence that the use of protoimperative signals by animals is a widespread phenomenon. When, for example, your cat vocalizes at you in the vicinity of the window and at the same time glances back and forth from the window to you, the cat is using a protoimperative signal that can be interpreted as "I want to go out." But it is very unlikely that your cat would use these same communicative signals to let you know that it has noticed something interesting in the garden.

Even though the studies with the language-trained Sherman, Austin, and Kanzi undoubtedly show that these apes can learn the symbolic function of a symbol, evidence that they can spontaneously use this cognitive ability outside the restricted context of experimental settings has not yet been provided. As will be seen in the next section, referential skills do not appear *de novo* in humans; instead, they are prepared for or promoted, prior to the child's acquisition of language, through a number of early communicative behaviors between competent adults and the infant, such as joint attention to objects (Bruner, 1983).

Predication, Grammar, and Triangularity

Further views on the differences between natural animal communication and human language are proposed by Bickerton (1990), who describes animal communication as *holistic*, concerned with the communication of whole situations. For example, the units of animal communication convey whole chunks of information. These chunks, as expressed for example in vervet alarm calls (see above and Chapter 5), are roughly equivalent to "A predator just appeared!" or "Look out! A lion's coming!" By contrast, human language deals mainly with entities—other creatures, objects, or ideas—to which states or actions are attributed: these are the predicates of the entities.

The flexibility of human language is illustrated by its use of different combinations of lexical items and of grammatical items in sentence construction. Thus, human syntax allows the expression of a different meaning merely by changing the order of the words in a sentence: "The truck follows the car" versus "The car follows the truck." But human syntax can also place the same words in different orders to convey the same meaning: "The cat eats the mouse" versus "The mouse is eaten by the cat." Finally, the same words placed in almost the same order can mean quite different things: "The woman that carried the child called the dog" versus "The woman carried the child that called the dog." Bickerton (1990), for whom animal communication differs qualitatively from human language, suggests that these changes to meaning are made possible because two categories of items intervene: grammatical items (articles, prepositions, auxiliaries, etc.) and lexical items (that is, those words which have an identified entity in the real world). Thus, the grammatical items *that* or *by* differ from the lexical items *woman* or *car* in that "the latter refer (if only indirectly) to some entity or class of entities in the real world, whereas the former do not really refer at all, but rather serve to express structural relations between items that do refer" (Bickerton, 1990, p. 10). Bickerton states that there is nothing in animal communication that corresponds to grammatical items and that, with very few exceptions (such as the *if/then* conditional used by Premack's chimpanzee Sarah), the vocabulary of apes, either spontaneous or taught, is strictly limited to lexical items.

One of the distinctive features of human language is the fact that words do not directly refer to entities or objects in the real world but to other words or signs forming a system. Thus, the relation between a symbol and an object is more than the simple correspondence between the two. Because the symbol is tied to a conception, we have a triangular connection among objects, symbols, and concepts: "It is the conceptions, not the things, that symbols directly mean" (Langer, cited in von Glaserfeld, 1977). The use of signs (either gestural or lexigrams) by reference to a system still awaits demonstration in ape communication. The appropriate use of lexigrams by chimpanzees does not allow the conclusion that these symbols are utilized in the same way humans use a linguistic sign, since their mastery of instrumental functions of these lexigrams in the test situations is sufficient for them to correctly adapt to these situations.

Some interesting secondary properties derive from this systemic organization of language (Bickerton, 1990). One such property deserves mention here: while animals communicate about things that have evolutionary significance for them (food, predators, sexual partners), humans can communicate about anything (e.g., in fairy tales or in science fiction), including entities that do not exist at all (this is the feature of productivity mentioned above).

The "Radical Arbitrariness" of Linguistic Signs

For some authors—such as Piaget (see Chapter 3)—language is one among several expressions of the semiotic function in humans, which consists mainly in the establishment of mental images. In symbolic play, for example, a mental image can be constructed involving two objects (a truck is represented by a matchbox, for example); from these two objects a single representation has emerged, with the consequence that each one can be substituted for another and with the condition that the objects share certain common characteristics.

This notion that an object (the *referent*) can be represented by another object (the *signifier*) has been borrowed from the work of the linguist Ferdinand de Saussure (1966). In language, the relation between referent and signifier is qualified as arbitrary, because there is no physical or analogical resemblance between the sequence of sounds and the content which is represented. In this respect, most of Washoe's gestures, Sarah's tokens, and the lexigrams operated by Austin, Sherman and other language-trained chimpanzees entertain an arbitrary relation with the various aspects of reality they represent. For de Saussure, however, the "radical arbitrariness" that characterizes verbal units is of a higher level of difficulty than the simple relation between two realities (see also Bickerton, 1990, and Vauclair, 1990b). In fact, two types of material reality need to be processed by the subject in order to comprehend or make a verbal sign: there is, on the one hand, the acoustic property of the sign and, on the other hand, the material property corresponding to the content expressed by the sign. In other words, a verbal sign is not simply a relation between material elements (sounds) and the content to which they refer (objects or actions). It is, rather, the product of two representations, one built on the acoustic material and another built on the meaning (conceptual image). The relation between the two images is said to be arbitrary because all natural languages have selected, through social convention in an arbitrary manner, a sequence of sounds to stand for a particular concept. It is precisely this conventional and arbitrary relation between a signifier and its referent that is called "radical arbitrariness." Although the construction of conceptual and acoustic images is typically an individual activity, the basic operation of language—that is, the *designation* or creation of signs—is nevertheless performed through social convention.

How can this analysis based on human languages help to clarify the issue of the linguistic nature of the chimpanzees' productions? In order to demonstrate that an ape uses signs that are equivalent to verbal signs, one should, from the present perspective, be able to show (1) that the ape possesses an individual representation of the signifier (of a gesture, for example) and of its content or meaning; (2) that a social convention has made the analysis of the representation possible, and (3) that the representation can be grasped by opposition to sets of other signs of the language.

These considerations have arisen from a conception of language at its most sophisticated level, as it is expressed by competent human adults, and it is not to be expected that linguistically trained chimpanzees should display these skills. The capacity to dissociate verbal signs from the reality they designate is not mastered by most human children acquiring language. Five-year-old children attribute specific meanings to words; these meanings are distinct from the usual meanings of their referents (Papandropoulou and Sinclair, 1974), indicating only a partial differentiation between the signifier and the referent. For instance, when 5-year-old children are asked to provide an example of a long word, they usually designate a long object (a train, perhaps). The ability to conceive that words can have an existence independently of the properties of their referents is mastered at around 7 years of age, but the comprehension that nouns are arbitrarily assigned to objects takes even longer to develop (Ianco-Worrall, 1972).

Pre-Linguistic Communication in Human Infants and Chimpanzee Infants

The preceding section has highlighted a number of differences between the communication skills of language-trained animals (marine mammals and apes) and the possibilities offered by the mastery of human language. Another way of comparing human skills with the abilities of different species is to consider the modes of exchange a human infant experiences during development, particularly during the period that precedes the onset of spoken language. Notwithstanding the infant's lack of a spoken language, the type and organization of social exchanges during this period seem to play a crucial role in preparing the infant for verbal communication later. It appears that these exchanges have no real equivalent in other primates.

The complex nature of prelinguistic interaction, notably between mother and infant, has been extensively investigated in the last twenty years. These interactions take, as one example, the form of mutual attention to an external object or event, which requires joint visual attention, pointing gestures, contact with objects, and referential language (Schaffer, 1984; Papousek and Papousek, 1987). Observations on chimpanzees (Goodall, 1968) suggest that mother-infant relations during the first months of life are mediated mostly through tactile contact rather than through vision. In this respect, early mother-infant interaction is different from what is observed in humans, for whom joint visual attention to objects is an essential part of the exchanges.

To illustrate the difference between humans and nonhuman primates with respect to communication during infancy, consider the following comparative study with three primate species: the human infant, the pygmy chimpanzee or bonobo, and the common chimpanzee (Vauclair and Bard, 1983). The purpose of the study is to discover the type and especially the levels of complexity of object manipulations occurring between 8 and 12 months of age and to analyze the communicative acts of the adult during object-oriented play. Subjects are given the same set of objects (four nested cubes, sticks, a plate, a cup). The human infant is observed in her home in the presence of the mother. The chimpanzee is observed in the nursery of the Yerkes Primate Research Center in Atlanta in the presence of an adult human caretaker, and the bonobo (Kanzi, see above) is observed with an adult female bonobo and a human caretaker.

Object manipulation is classified according to a broadly Piagetian system designed to record behaviors ranging from simple visual orientation toward objects, to secondary circular reactions and their coordination, and, finally, to conventional use of objects (Piaget, 1952a, 1954). The study found differences in the quality and quantity of manipulations by the human infant and the apes. For example, the human infant usually held the objects above the ground. Moreover, they often transferred objects from one hand to the other. These activities are rarely observed in the ape infants, who typically simply held the object or moved it against a substrate (Vauclair and Bard, 1983). These findings are confirmed by other studies, on gorillas (Redshaw, 1978), on chimpanzees (Mathieu and Bergeron, 1981; Mignault, 1985; Poti and Spinozzi, 1994), and on Japanese macaques (Torigoe, 1986).

In order to analyze the communicative behavior of the adults in relation to the infants' object manipulations, Bard and Vauclair (1984) asked: (1) whether adults act on objects so as to engage the infant's attention with those objects, and (2) whether object manipulations by adults influence the infants' behavior with the objects. Analyses of the sessions just described (Vauclair and Bard, 1983) indicate that adult apes rarely act on objects with the apparent intent of engaging the infant's attention, whereas adult humans manipulate objects primarily with the intent of stimulating, sustaining, or enhancing the infant's actions on the objects. Infant apes respond differentially; although they do not attend to the manipulation of objects by adult apes, one of them attends to and even manipulates objects when interacting with an adult human. More specifically, the bonobo infant typically does not attend to the adult bonobo's actions. When he attends, the mother is acting neutrally. This bonobo infant shows similar object-oriented responses to the acts of both the bonobo mother and the human caretaker, although these two adults act differently. The human often attempts to engage the infant's attention with objects, but the bonobo infant typically does not attend to the adult's actions. The common chimpanzee infant, when in the presence of a human caretaker, expresses object-oriented responses that are very similar to those observed in the human infant, including high frequencies of appropriate responses (such as paying attention) to the adult's attempts to engage the infant, and a high proportion of instances in which the infant does not attend when the adult is acting neutrally.

Additional data confirm that human-reared chimpanzee infants show some early behavioral patterns more similar to those of humans than to those of mother-reared conspecifics (Bard, Platzman, Lester, and Suomi, 1992). Russon (1990) further discusses monkey-ape-human differences concerning the role of objects in interactions between peers, and relates the species differences to findings from Piagetian studies.

Modes of Exchange in Human and Nonhuman Primates

In general, what makes humans remarkable is the early mutual exchange that takes place between adults and infants regarding inanimate objects. This feature, along with other aspects of human ontogeny (for example, retarded locomotion), brings about a specific mode of contact with the external world which can best be described by a general ability to use intermediaries to control both social and physical objects (Vauclair, 1984; see also Spinozzi and Natale, 1986). It has been shown by Bates (1979) that the use of objects as instruments and the combinatorial play involving secondary and tertiary circular reactions are good predictors of gestural and verbal communicative development in young children. These findings have led this author to speculate that there is a common cognitive "software" underlying both capacities and that this common mechanism is best characterized as the capacity to understand means-end relations (see also Chapter 4). Just as the emergence of communicative intentionality implies that the infant can differentiate the communicative gesture from the purpose of that act, so must the use of an object as a tool to accomplish a particular goal become differentiated from the means of acting upon that object. Both tool use and communicative gestures require an intermediary means to accomplish a goal. When a child looks at an adult and gestures toward an object that is out of reach, he is engaging in a cognitive activity that is very similar to that of using a string to pull an object closer.

In fact, by the end of the first year, the human primate uses three kinds of tools: inanimate objects, human agents to operate on objects, and the object itself to operate on human attention (Bates et al., 1975). Although human infants show the clearest instances of socially mediated manipulations with objects, there are indications that great apes may approach similar complexity. For example, Plooij (1978) reports that at around 9 months of age wild infant chimpanzees start to use the mother as an agent:

they alternate between looking into her eyes and looking at her hand or mouth while reaching for a desired object, usually food held by the mother (see also Bard, 1990, for a similar example in orangutans). Another example is Gomez's (1991) description of a juvenile gorilla using a human as an agent to operate on an inanimate object (a door), by alternating its gaze in different combinations and by physically manipulating the agent. Of course, these examples in apes concern situations in which the infant desires an object, whereas in humans the resulting joint attention to the object may be a goal in itself (Bruner, 1983). The available information on the interface between the social environment and object manipulation and tool use in monkeys, apes, and humans is summarized in Vauclair and Anderson (1994).

It is likely that in chimpanzees interactive structures involve vocal, gestural, and emotional exchanges but not inanimate objects (with the notable exception of food). This feature could explain why chimpanzee mothers do not engage their infants in the manipulation of inanimate objects. Altogether, the world of objects, space and time are presented to human infants in a socially structured way. The various skills developed by the human infant (referential communication, tool use, complex object manipulation) reflect the shared meanings or understandings that structure the context in which language develops.

Summary and Current Debate

A perennial question in the field of animal cognition is the degree to which animal communication resembles the most elaborate system that we, as humans, use to communicate—namely, vocal language. Several avenues exist for comparing animal communicative systems and the human structure of language. One of them is simply to list the basic traits of language and search for corresponding features in animal communication. This approach is supplemented by other, more direct comparisons, exemplified by attempts to teach marine mammals and apes some of the rudiments of human language. This research offers a stimulating framework for discussing differences and resemblances in the use of signs between apes and children, and it has shown that communication systems in various primate species, although they share some common features, differ in some important aspects.

The main differences are structural and functional. One structural difference concerns, for example, the fact that human language is organized in a two-level system, what is called a duality of patterning. This duality is not limited to verbal communication, since it can also apply to all forms of gestural expression and communication. As an example, the very simple gesture of indicating assent (yes) appears to be included into a structure of conventional oppositions; its meaning does not arise from the motor acts actually expressed but from opposition to other gestures or attitudes that could occur in the same context, such as those of refusal or of motionless and silent attitudes or even those expressing perplexity. In other words, the structure of gestural expressions implies the same duality of patterning as that described for linguistic signs.

With respect to functional features of human communicatory systems, this chapter has advanced the position that communication in animals has essentially an imperative function (as illustrated by the language-trained chimpanzees using learned symbols to make requests). Humans, on the other hand, use linguistic signs for both imperative and declarative purposes (for example, two persons sharing an interest toward a third person, object, or event).

It is important to realize that the possession of language with all its properties endows the human subject with a structure that requires interplay between at least three main operations. First, the subject must detach the representations and the communicative expressions from the stimuli and signals that he either receives or emits in his physical and social surroundings. Second, the subject must combine these representations and expressions within a structure in such a way that each of them represents and/or expresses what the others do not represent and/or express. Third, the subject must address these expressions to a partner or relay them to an object, in a way that takes into account the triadic relationship in which he is involved with the two other partners. Possession of this triadic structure allows the human subject to form a number of relationships unmatched by animal species (Vauclair and Vidal, 1994). Two will serve as examples here.

First, the symbolic and pretend-play activities described by numerous psychologists (e.g., Piaget, see Chapter 3) allow the child not only to play with objects as if they were other objects but also to treat them as companions. (This has been discussed in terms of the "transitional object" by Winnicott, 1971.) Pretending in this way implies the detachment of the object from its immediate function in order to use it in a sequence of symbolic and arbitrary acts. It is noteworthy that object use of comparable complexity has not been described for nonhuman species. Only a few cases of pretend plays have been reported (Parker and Milbrath, 1994), and these only in great apes reared by humans. For example, Vicki, the chimpanzee raised by Hayes (1951), was observed pulling an invisible string on an invisible toy.

Second, communicative behaviors in animals are apparently performed within dyadic (and not triadic) systems of interaction. This phenomenon appears to be the rule, for example, in social relations. In Chapter 5 it was noted that vervet monkey mothers are able to recognize the vocalizations of their offspring (Cheney and Seyfarth, 1980). These behaviors have certain limitations, however, that are unlikely to occur in humans: the onlooking females maintain their distance with respect to both the mother and the young, for instance, and they do not appear to attempt to reestablish the relation between the mother and child involved.

Situations in which one animal does appear to show attention to a triadic relationship involving itself, another animal, and some third party appear in fact to be remarkably rare. A few cases of "prototriadic structures" are observed in some social relationships of nonhuman primates, however. A male baboon may refrain from courting a familiar female he has observe to be already paired with a familiar male. Nevertheless, its restraint is only short-lived: it disappears when the paired male is out of sight or when the female remains alone for some time (Kummer, Götz, and Angst, 1974). De Waal (1982) provides examples of the involvement of a third individual in the formation of coalitions by chimpanzees; he has called this phenomenon "triadic awareness." It seems that even when nonhuman primates exhibit attention to relationships between at least two other individuals, they do not appear to engage in symbolic triadic relationships. In other words, contrary to humans, animals do not seem to confront the *other* of the other.

The use of declaratives and the capacity for joint attention to objects, described earlier in this chapter, could be the antecedents of these possibly unique features of human language, and they could set a framework that allows the development of mental attribution of beliefs, knowledge, desires, and intentions to social partners. After all, spoken declaratives constitute an elaborate form of joint attention, by which a given speaker attempts to affect the listener's mind (Baron-Cohen, 1992). In this same line of thinking, protodeclarative and declarative behaviors may be precursors to the development of a "theory of mind." This question, posed in relation to the attribution of mental states, will be dealt with in the next chapter.

7

Imitation, Self-Recognition, and the Theory of Mind

A major issue in the field of animal cognition concerns the acquisition of behaviors by individuals, either young or adult, in the wild or in captivity. Are behaviors acquired through social devices such as imitation, or must each individual confronted with a novel situation "reinvent" the appropriate behavior for itself? To carry the question further, can behaviors become traditional, or even cultural, within the group or population in which they have been established?

Other questions to be considered in this chapter concern the formation by animals of knowledge about themselves and about others' selves. An enquiry into self-knowledge amounts to asking whether a given animal has some control over its own cognitive processes, a sense of its identity (for example, of its physical appearance), or even some sort of self-awareness. Knowledge of others can be expressed in different ways, such as the ability to control the behaviors of conspecifics through deception.

As observed by Premack (1985), traditional approaches to animal learning and cognition have raised questions of *instantiation:* "Do rats have representations?" "Do chimpanzees have intentions?" To amass new kinds of evidence going beyond what an animal does, to investigate what it "thinks" about itself and about others, we need a new paradigm: *attribution.* We must ask, for example, "Do chimpanzees think that others have intentions or beliefs?" This line of enquiry is also known, since the work of Premack and Woodruff (1978), as the search for a "theory of mind." The problem of attributing beliefs, desires, or intentions to animals is somewhat similar to the problem of language, because language functions as a system of intentions: language implies that the sender is able to manipulate its message in order to control the behavior of the receiver. But, of course, the medium for conveying intentions does not need to be linguistic, and so the search for intentions might be more amenable to experimental investigations with animals than is the search for language equivalences.

Is There Evidence for Imitation in Animals?

The answer to this question is far from decisive. Controversies arise with respect to the role played by social factors in the establishment of individual behaviors, and there are also disagreements over the definition of concepts such as imitation.

The question of how animals acquire behaviors is central to the comparative analysis of cognition. The first possibility to be considered is that a given animal learns by imitating actions performed by competent conspecifics. Empirical studies of imitation in animals began at the dawn of experimental animal psychology, most notably in the works of Romanes, Morgan, and Thorndike (Galef, 1988; see also Chapter 1). It is impossible from the available literature to reach a consensus for an operational definition of imitation. One researcher (Galef, 1988) has listed 22 terms for imitative behavior. I prefer to make a simple distinction between "imitation" and "social learning" (Heyes, 1993a). Imitation will be defined as *learning about behavior* through the observation of or interaction with conspecifics. By contrast, in social learning, observers *learn about stimuli,* objects or events in the environment, either how to distinguish them from other classes or whether to attach a positive or negative value to them by virtue of their relationships with other objects and events (Heyes, 1993a).

Cases of social learning are numerous in animal life. It is observed in vertebrates, but also in some invertebrates, for example, in approach behaviors (toward food, for example) as well as in avoidance reactions (away from predators). One case of learning by observation has been reported for octopuses (Fiorito and Scotto, 1992). The task for the octopus is to attack a red ball in a tank and neglect a white ball. After the model subject has been trained, a naive octopus (the observer) is allowed to watch the demonstrator's performance through a transparent wall. The observers consistently select the same object as the demonstrators. Moreover, observers learn more rapidly than the demonstrators, suggesting that social learning is more efficient than individual training. As pointed out by Suboski, Muir, and Hall (1993), the behavior of observer octopuses cannot be properly called "imitation-by-copying," because although the observers might have learned *which* stimulus to respond to, they have not learned *how* to respond to the stimulus. The behaviors of the octopuses thus constitute typical acts that can be acquired through social learning, but they do not satisfy the definition of imitative learning, in which the observers "acquire, as a result of observing a conspecific's behavior, X, the capacity to execute a behavior that is topographically similar to X" (Heyes, 1993a, p. 1000).

It is striking, and somewhat paradoxical, to note that the evidence for imitation is stronger for birds and some small mammals than for primates (Galef, 1988; Visalberghi and Fragaszy, 1990; but see Russon and Galdikas, 1993). Imitative behavior, which is sometimes called "copying," can be found in pigeons, for example. In one study (Palameta and Lefebvre, 1985), an observer pigeon can see a trained conspecific piercing a hole in a sheet of paper covering a box of seeds and then eating some of the food inside. When the observers are presented with a similar food box, they learn to feed from paper-covered food boxes more rapidly than the control pigeons, which have observed piercing alone, eating alone, or neither piercing nor eating. A similar phenomenon has been studied in budgerigars (Dawson and Foss, 1965; Galef, Manzig, and Field, 1986): observer birds tend to use the same appendage (feet or beak) as demonstrators to remove a flat cover from the top of a food cup. Cases of imitation between persons and a parrot (the African grey parrot) were reported in a study by Moore (1993). The bird, which was shown a few stereotyped actions each day (such as head shaking), spontaneously imitated the movements made by the human and accompanied them with words or phrases as labels.

In the same vein, evidence of imitation is available in studies with rats (Heyes and Dawson, 1990; Heyes, Dawson, and Nokes, 1992; Heyes, Jaldow, Nokes, and Dawson, 1994) using a "bidirectional control" procedure. This procedure allows the investigator to distinguish between imitation and a social learning process other than imitation—namely, the control of cues provided by conspecifics (this kind of social learning is also called "social enhancement" by Thorpe, 1956). Social or stimulus enhancement occurs when a behaving conspecific directs the subject animal's attention to a particular object or to a particular part of the environment. In Heyes's experiments, rats observe a trained conspecific pushing a joystick that is suspended from the ceiling of one compartment of a duplicate cage (see Figure 7.1). The joystick is moved either to the right or to the left. Half of the animals observe the joystick moving automatically, and half observe a conspecific manipulating the joystick. Moving the joystick in the correct direction is food-reinforced. Rats that observe a demonstrator produce a response bias that is congruent with the observed direction of joystick movement. Rats that observe the joystick moving automatically, either in the presence or absence of a passive conspecific, do not have a consistent directional preference. According to the authors, the observer rats' bias in responding like the demonstrators is not due to stimulus enhancement but to response learning by observation, in other words, imitation.

Imitation has been documented in captive dolphins (Tayler and Saayman, 1973). A well-known example concerns the case of an infant dolphin that saw a human observer blowing smoke against the glass of the tank. The infant swam off to its mother, returned, and released a mouthful of milk, apparently to mimic the human's puff of cigarette smoke.

A number of studies are available on imitation by great apes. Cases of the imitation of motor acts were reported for the home-raised chimpanzee Vicki (Hayes and Hayes, 1952) as well as for sign-language-trained chimpanzees (Gardner et al., 1989). Studies with captive chimpanzees also demonstrate that chimpanzees have some ability to imitate human gestures (Custance and Bard, 1994) and tool use (see below; Tomasello, Davis-Dasilva, Camak, and Bard, 1987). Finally, the very few attempts that

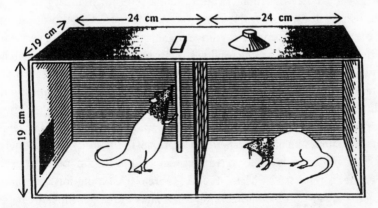

Figure 7.1. The joystick apparatus used to test imitation in the rat. (From Heyes and Dawson, 1990; reprinted with permission from Academic Press Ltd.)

have been made to train monkeys to imitate have failed to reach positive evidence. For example, Mitchell and Anderson (1993) trained an adult male long-tailed macaque to imitate the scratching behavior of a model (the experimenter). This monkey did learn to scratch when the model scratched, but he never learned to generalize the behavior, as when the model scratched new target areas.

Imitation and Cognitive Processes

The ability of an animal to copy a novel act when the act is no longer visible implies that the animal is able to internally represent the observed act. In this respect, the agenda of experimental studies of animal cognition is replete with examples showing that, indeed, animals of different phyla are able to code, store, and retrieve information in order to produce an adapted response (see, for example, Chapter 1).

For Piaget, the capacity for imitation develops in human infants through sensorimotor sequences similar to those responsible for object concept, spatial organization, and causality (see Chapter 3). Piaget (1962) claimed that the capacity for real imitation (called "deferred imitation")—namely, imitation of the actions of a model—appears quite late in human infants (between 12 and 18 months of age). This aptitude for imitation reveals that the infant has the capacity for mentally representing the actions (gestures, but also words) performed by a partner. Piaget's view that this aptitude is acquired in late infancy has been strongly challenged by later research with human neonates. Meltzoff and Moore (1977) demonstrated that 12- to 21-day-old infants can imitate a series of manual gestures or facial expressions, such as sticking out the tongue. For those authors, infants possess a perceptual-cognitive organization of high level very early in life (if not at birth), which allows them to detect intermodal matches (between vision and proprioception), expressed in these early forms of imitation. From this point of view, it could be that such a supramodal ability differs from mental representation in the Piagetian sense. In effect, the problems of symbolic imitation, supramodality, and mental representation should be distinguished. For example, in the study with neonates, symbolic imitation is confounded with deferred imitation, because in the experiments reported by Meltzoff and Moore (1977) imitation is observed immediately after the demonstrator terminates the target movement.

One can therefore question whether the imitation performed by neonates is true symbolic imitation. Is the imitation of tongue protrusion identical to the symbolic representation reported by Piaget (1962, p. 66) in an infant imitating the actions of inanimate objects (here a box of matches) in several ways (opening and closing, reproducing the sound of the box)? Sticking out the tongue is a deferred motor response involving some form of short-term memory, whereas symbolic imitation implies the mental reconstruction of an entire motor sequence.

Returning to the experimental evidence gathered so far regarding imitation in animals, it is quite clear that the tasks that have been studied in animal subjects do not require any symbolic treatment of the kind involved in deferred imitation studies in human infants. However, imitative capacities shown by rats and other species indicate that these imitations imply some form of mapping between a visual input coming from the model and some kinesthetic or proprioceptive stimulations from the animals' own actions (Heyes, 1993a). This set of abilities is called cross-modal performance (i.e., the ability to equate patterns of sensory stimulation in different modalities). At a minimum, however, imitation requires that the subject construct some sort of supramodal or amodal representation of the act it has seen performed to reproduce it successfully during the imitative action.

Tradition and Culture

The authors of some recent studies (Galef, 1992; Tomasello, 1990; Tomasello, Kruger, and Ratner, 1993; most data concern nonhuman primates) have argued that animals acquire new behaviors not by reproducing the behavior of conspecifics but by inventing them themselves. In this sense, the behaviors do not deserve to be labeled "traditional" or "cultural." In contrast, other authors (McGrew, 1992; Boesch, 1993a,b) conclude from studies of wild chimpanzees that modifications of behaviors accumulate over generations. There is even evidence that some behaviors can result from purely social conventions.

A "tradition" is an activity that is learned in some way from others and then passed on to naive individuals; "culture" requires, in addition to these criteria, the intervention of a teaching act. Several cases of "protoculture" have been described by primatologists. One such case concerns the washing of sweet potatoes by Japanese macaques on the Island of Koshima (Kawai, 1965; see also the section on tool use in Chapter 4). The first instance of this behavior was observed in 1953 in an 18-month-old female. This monkey began "to take pieces of sweet potato covered with sand to

a stream and to wash the sand from the potato pieces before eating them"
(Galef, 1992, pp. 162–163). This behavior eventually spread through the
Koshima population of macaques, following the lines of matrilineal affiliation. From an analysis of several factors (including the slow propagation
of the behavior and the provision of food by the human observers) in
relation to the apparent invention and spread of this novel behavior, Galef
(1990, 1992) has come to the conclusion that potato washing is unlikely
to have been acquired through imitation. For Galef, it is best described as
a local-specific behavior and not the result of a tradition.

In contrast, some behaviors involving the manipulation of plant material have been shown to expand rather rapidly among groups or populations of chimpanzees. This is the case for a behavior known as leaf-clipping. "In leaf-clipping, the performing chimpanzee noisily pulls to bits one
or more leaves by hand and mouth, leaving only the stripped petiole"
(McGrew, 1993, p. 188). According to Boesch (1993a), the function of
leaf-clipping varies among different populations of chimpanzees: it is part
of a courtship display in Mahale chimpanzees (Tanzania), but it serves as
a signal of frustration among the chimpanzees of the Bossou community
(Guinea). Moreover, Boesch describes a rapid (within a month) and novel
use of leaf-clipping among most of the chimpanzees of the Taï Forest
(Ivory Coast).

Questions of tradition and culture in relation to animal behavior gen-
erate debates and controversies. The issue is far from being resolved, given
that stringent conditions must be met if the traditional nature of a given
behavior is to be demonstrated. As Galef (1990) has suggested, both field
observations and experimentation are necessary for properly assessing the
traditional status of a behavior. Traditional or "local-specific" patterns of
behaviors can be discovered only through careful observations of the
behavior in many different populations. After a behavior has been identified as a possible tradition, experimentation in controlled conditions may
be needed in order to "determine whether a local-specific behavior observed in the field can be acquired by social learning and should be
considered traditional" (Galef, 1990, p. 91).

The Attribution of Mental States in Animals

Only recently has the question of mental states in animals been addressed
in a systematic way, including the use of laboratory experiments. The
following sections will examine several attempts (mostly with apes) to

understand the extent and the limits of mental attributions in animals. Premack and Woodruff (1978) conducted a number of original experiments on chimpanzees' comprehension and attribution of mental states during social exchanges with human partners. As they explain in a picturesque way, they are less interested in the "ape as a physicist than as a psychologist." In other words, the focus of interest is less the chimpanzee's comprehension of physical relations during tool-use activities (see Chapter 4) than how its knowledge "about the physical world affects what he knows about what someone else knows" (Premack and Woodruff, 1978, p. 515).

Saying that an individual can mentally impute intentions, desires, and beliefs to himself or to conspecifics amounts to saying that this individual has a "theory of mind." A system of inferences about mental states can be viewed as a theory on two grounds, "first, because such states are not directly observable, and second, because the system can be used to make predictions, specifically about the behavior of other organisms" (Premack and Woodruff, 1978, p. 515).

Before examining how this theory of mind is experimentally tested, I will first introduce some spontaneous instances of communicative behaviors in birds and in nonhuman primates. These are of interest because they are suggestive of a certain amount of control over the information that circulates between individuals.

The Manipulation of Information during Communicative Exchanges

Manipulation of information during social exchanges has been described in several animal species, particularly birds. The previous chapter reported cases of male domestic chickens controlling their rate of food calling according to their food preferences and the presence of females (Marler et al., 1986a). In a subsequent experiment (Marler, Dufty, and Pickert, 1986b), the chicken is alternately presented with a highly prized food item (a mealworm) and with a non-food item (a fragment of nutshell). The food item is known to elicit food-calling behavior in the male. The two kinds of items are presented to the male chicken (the "sender" of food calls) in the presence of three different types of birds (the "receivers"): a strange hen, a familiar hen (the male's cage mate), or another adult male. Finally, in control trials, there is no audience at all. The frequency of the sender's vocalizations is analyzed both as a function of the type of receiver and of the nature of the item (food or non-food).

The results (see Figure 7.2) indicate that food calls are most frequent when females (familiar or not) are present; no food call is recorded in the

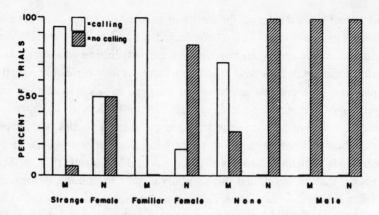

Figure 7.2. Results of a test of intentional communication in chickens. Percentages of trials in which male chickens either made *(open bars)* or did not make *(cross-hatched bars)* food calls with either a strange or a familiar female as audience, with no audience at all, or with another male as audience. In each case, data are given for calls made in the presence of a preferred food (mealworm = *M*) and a non-food item (nutshell = *N*). (From Marler et al., 1986b; reprinted with permission from Academic Press Ltd.)

presence of another male. Male chickens thus appear to control their vocalizations, taking into account the nature of their partner. More specifically, these calls are inhibited when the receiver is not a sexually appropriate partner. When confronted with non-food items, the chickens emit calls at a low frequency or refrain from calling. There is, however, a noticeable exception to this rule: when a non-food item is presented, the subjects produce significantly more calls in the presence of a strange female (on 50 percent of the trials) than in the presence of a familiar female (17 percent of the trials). The authors interpret this behavior as an *intention to misinform*, on the assumption that a male chicken is highly motivated to attract and establish a social bond with the unfamiliar hen.

In more natural contexts, some birds engage in "false" antipredator behaviors to distract intruders from approaching their offspring. This is, for example, the case of the so-called broken-wing display performed by piping plovers (Ristau, 1991a). These birds flap their wings in a very awkward way, which makes them appear to be injured. This behavior effectively serves to draw the attention of a potential predator away from the nest and toward the "injured" bird.

Concealment of information and deceptive behavior. Numerous anecdotes are available suggesting that nonhuman primates manipulate information given to conspecifics (for a review, see Whiten and Byrne, 1988). These situations include cases of concealment of information from others and tactical deception. Kummer (1982) offers an example of a hiding tactic shown by a young male hamadryas baboon. Social organization in this species is characterized by the harem, which is composed of a single adult male and several females. In principle, the females within the harem mate only with the male leader, but field observations tell a different story: "A juvenile female hamadryas baboon in estrus leaves her adult male leader and repeatedly mates with a juvenile male behind a rock where her leader cannot see her. Between matings, she goes to where she can peek at the leader, or even approaches him and presents herself to him before she again mates with the juvenile in the hiding place" (Kummer, 1982, p. 118).

This observation is a very good example of the kind of unstructured field report from which one might infer some kind of attribution of knowledge on the part of an animal. According to the mentalistic hypothesis, the juvenile male realizes that, because his behavior was not observed by the leader, the leader possesses knowledge different from his own. Thus, the subordinate male was able to copulate behind the rock because he knew that he could influence what the leader perceived and therefore knew. An explanation in terms of mental attribution might, however, be challenged by a simpler one, based on the baboon's prior experience of such situations. From this perspective, the young male may have learned that whenever he copulates out of sight of the dominant male, he also avoids being threatened by him. A type of information concealment occurs in parallel in the female, too. Observers note that the juvenile female's behavior is a case of "acoustic hiding," in that she suppresses the vocalizations usually associated with mating. This acoustic concealment may be the result of an active manipulation to fool the dominant male (strong hypothesis), or it may be due to aversive conditioning (weak hypothesis), in which the female inhibits the vocalizations to avoid being punished by the dominant male.

Deception has been studied experimentally in chimpanzees (Woodruff and Premack, 1979). First, an operational definition of deception requires that the sender understands and controls the effects that his actions can have on the recipient. An experiment was designed to evaluate the capacities of the sender to modulate the information he transmits, either by altering the information or by suppressing it. Both the ability to deceive and susceptibility to being deceived were explored.

In the experiment, a human and a chimpanzee communicate about the location of a food item, which is hidden in one of two containers placed on the floor of the experimental room. At the start of each trial, the chimpanzee observes the baiting of one container, but the containers are inaccessible to the animal. After baiting, a trainer enters the room and the chimpanzee can indicate to this partner the location of the hidden food. The chimpanzee is confronted with two kinds of trainers, a cooperative trainer and a competitive trainer. Both trainers are distinguishable by the way they are dressed and by their behavior: the cooperative trainer wears a green coat and is friendly toward the chimpanzee; the competitive trainer wears black boots, a white coat, dark sunglasses, and a cloth over his mouth and behaves in a brusque manner. Above all, if the cooperative trainer happens to retrieve the hidden food by following the directions (orientation or gestures) given by the chimpanzee, he will share the food with the chimpanzee. By contrast, the competitive trainer will keep all the food for himself. Four juvenile chimpanzees are tested in a *production task*, whereby they indicate the location of the food to the trainer, and in a *comprehension task*, whereby they use information from the informed human to determine which of the two containers is baited. In the latter task, the cooperative trainer always orients toward the baited container, whereas the competitive trainer always moves toward the unbaited one.

It was found that the cooperative trainer can rapidly (after about 120 trials) interpret the behavior of the chimpanzee and find the food. The chimpanzee typically moves toward the correct container, extending one leg toward it, and then glances back and forth from trainer to container. The behavior of the chimpanzees is similar toward competitor and cooperative trainer, at least at the outset of the test. But with the repetition of trials, the subjects learned to suppress their behavior in the presence of the competitor. One of the chimpanzees even began to deliberately misinform the competitor, by pointing with the foot to the incorrect container. In the comprehension task, three out of four chimpanzees are able to use the cues provided by the cooperative trainer and to orient toward the correct container. In response to the competitive trainer, all chimpanzees began by following the (wrong) directions. Over the course of the experiment, three of the four subjects learned to controvert the competitive trainer's cues.

For Woodruff and Premack (1979), the flexibility demonstrated by the chimpanzees in the manipulation of the information they receive or send suggests the existence of a real capacity for deception in this species.

This experiment, although unique in its kind, has been criticized on a number of grounds, and it is questioned whether it demonstrates deception in the chimpanzee. One criticism is that the chimpanzees have simply learned what behavior is required to obtain the food reward, namely, respond in one fashion with the cooperative trainer and in another fashion with the competitor (Ristau, 1986). To extend this criticism, one may wonder whether the chimpanzees have really attributed sets of mental states or whether they have merely reacted to differences in appearance or behavior. In other words, the basic problem is to be sure that the chimpanzees are acting on the basis of reasoning about observables (the "RO" interpretation), or if they reason about mental states (the "RAM" interpretation: Heyes, 1993b). As a step toward choosing between these interpretations, certain control experiments have been proposed. Thus, the food could be placed in plastic, see-through containers and the competitive trainer could be instructed to conspicuously look at these containers when he enters the room. "If the chimpanzee can instantly recognize a logical inconsistency between (a) what it wants the trainer to believe and (b) what the trainer obviously knows, then we would have some stronger evidence for the notion of conscious deceit" (Griffin, 1982, p. 397).

Guessing and knowing. When the term is applied to humans, *mental attribution* refers to various categories of behavioral expressions, including beliefs, desires, fantasies, mental entities, and private selves (Wellman, 1990). A rudimentary form of social attribution concerns the understanding of the causal connection between visual perception and knowledge formation. This problem—namely, how an individual can understand the consequences of the visual experiences of others—has been investigated in chimpanzees, in macaques and in human infants. In the study with chimpanzees (Povinelli, Nelson, and Boysen, 1990), the subjects watch two experimenters performing two roles, which will be referred to as the "knower" and the "guesser." The chimpanzee identifies the location of hidden food according to directions from the two experimenters, who alternate between the two roles. Prior to the experiment, all subjects have been trained to point to baited food cups and to respond (pull a handle to bring in the food cup) to pointing by a trainer. At the start of each trial, the knower baits one of 4 obscured cups while the subject watches the process but cannot see which of the cups contains the food (the guesser waits outside the room until the food is hidden).

After the baiting, the guesser enters the room and both experimenters point at a container (the knower points at the baited container, the guesser at one of the other three). Two of the four subjects quickly learned (after 150 trials) to respond to the information provided by the knower and not the guesser. The two other subjects needed more trials to differentiate between the two experimenters (see Figure 7.3).

In order to test for the possibility that the chimpanzees simply select the container indicated by the experimenter who does the baiting or who is in the room at the time of baiting, a transfer experiment is performed. In this test, the baiting is done by a third experimenter, in the presence of both the knower and the guesser, but during the baiting the guesser covers his head with a paper bag while the knower watches the baiting. This transfer test has no effect on three of the four subjects, who maintain the

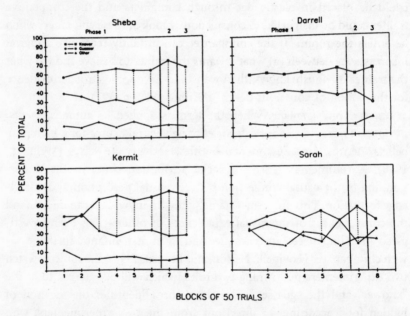

BLOCKS OF 50 TRIALS

Figure 7.3. Results of a test of perspective-taking abilities in chimpanzees. Perform-ance by four subjects is plotted across the different phases of the experiment. *Phase 1:* The guesser remains outside the room while the knower hides the food; at the end of phase 1, the knower wears a small blue hat. *Phase 2:* The hat is eliminated from the procedure. *Phase 3 (transfer test):* the guesser places a bag over the head while the knower watches a third experimenter hide the food. Vertical dotted lines indicate the blocks of trials in which the subjects show a statistically significant preference. Blocks of 50 trials were presented for Phases 1 and 2, Subjects complete 10 trials each day. (From Povinelli et al., 1990.)

discrimination between the two experimenters and continue to follow the knower's indications (see Figure 7.3).

The authors conclude that the chimpanzees, under the conditions tested, appear to know that someone who has witnessed an event has an understanding of the event that is different from the understanding of someone who does not have access to knowledge about the event. The chimpanzees have shown that they can understand the perception-knowledge relationship.

Four rhesus macaques received the same training as the chimpanzees and are tested in a similar experimental setup (Povinelli, Parks, and Novak, 1991). Instead of four cups, three containers are used. In contrast to the chimpanzee results, none of the four macaques exhibited a preference for the knower over the guesser, even after 12 to 16 weeks of testing (see Figure 7.4). The investigators came to the conclusion that rhesus monkeys, unlike chimpanzees, do not understand the seeing-knowing relationship (see also Povinelli, 1993).

For comparative purposes, this relation between seeing and knowing has been studied for groups of 3- and 4-year-old children (Povinelli and De Blois, 1992). Subjects have to discriminate between two adults, one who hides a surprise under one of four obscured cups, and one who leaves the room before the surprise is concealed. Most 4-year-olds (but not 3-year-olds) are able to discriminate between the two experimenters. Additionally, when the children are asked to explain how they know the location of the correct cup, only 4-year-olds can explain that the source of their knowledge is the experimenter who saw the surprise hidden. There is a difference between chimpanzees and humans in the performance of this task, however. All successful children need only ten trials or less to solve the task, while the quickest chimpanzees (two out of four) seem to understand the problem only after 150 trials. Nevertheless, this study with humans shows that the results of the chimpanzees are not trivial and suggests that the chimpanzees' ability to understand the distinction between guessing and knowing reflects a more general ability to model the visual perspectives of others. Note, however, that Gagliardi et al. (1995) have shown that the seeing and knowing task could be solved by humans via conditional discrimination, disclaiming thus the unique interpretation in terms of attributional abilities.

A naturalistic study of mental attribution. The previous experiments may appear rather artificial. In addition to involving social exchange between

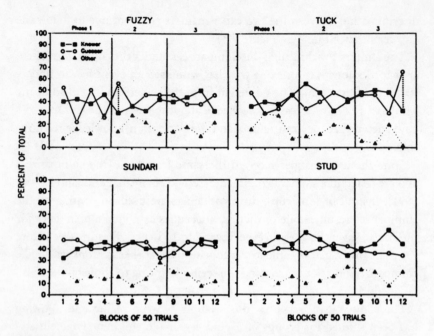

Figure 7.4. Results of a test of perspective-taking abilities in rhesus macaques; data given for the first three phases for all subjects. *Phase 1:* The guesser remains outside the room while the knower hides the food. *Phase 2:* The procedure is altered by having the knower wear a large red hat. *Phase 3:* The guesser places a bag over the head while the knower watches a third experimenter hide the food. (From Povinelli et al., 1991.)

two species, a chimpanzee or a macaque and a human experimenter, they are removed from the spontaneous repertoire of primates and from the real challenges of their social life. Cheney and Seyfarth (1990c) arranged a sort of naturalistic experiment with macaques to study the monkeys' capacities for distinguishing between their own knowledge and the knowledge of others. Macaques frequently utter alarm calls in the face of a potential danger (e.g., when they see a technician holding a net used to capture the monkeys). They also emit food calls when they discover one of their preferred foods.

Adult females and their juvenile offspring are shown two scenes: in one scene, some preferred food is placed by a human in a bin in the test area; in the second scene, mimicking a predator-like situation, a technician wearing a surgical mask and brandishing a capture net hinds behind a barrier in the testing environment. Two conditions are tested. In the "knowledgeable" condition, both mother and offspring are seated together and can watch the two scenes; after observing the two scenes, the offspring, but not the mother, is released in the arena. In the "ignorant" condition, the offspring is close to the mother but is visually isolated from her by a steel partition, such that only the mother can see the two scenes. The object of this experiment is to determine if the mother will attempt to alert her ignorant offspring more often than she attempts to alert a knowledgeable one. The results show that the mother does not behave differently under the two conditions. In particular, mothers did not emit alarm calls to "warn" ignorant offspring about the presence of danger.

Cheney and Seyfarth conclude that the absence of differential reactions from the mothers indicates their inability to take into account the knowledge or ignorance of their infants and thus to attribute mental states.

Criticisms of the techniques and interpretations of the above experiments, and more generally of the possibility of mental attribution in primates, can be found in the commentaries following the articles by Premack and Woodruff (1978), Whiten and Byrne (1988), and Cheney and Seyfarth (1992). The main point of the criticisms that have been raised can be summarized with the distinction proposed above between reasoning by observables (the "RO" interpretation) and reasoning by mental states (the "RAM" interpretation). It seems that in most experiments on mental attribution, given the training methods and the test situations, one cannot discount the possibility of an RO interpretation in favor of an RAM interpretation. Systematic transfer tests are called for (Heyes, 1993b).

Self-Knowledge and Self-Recognition

A question of considerable interest today is the relation between social attribution, as evidenced by chimpanzees but not macaques, and these species' respective abilities for self-recognition and awareness of subjective states.

When faced with their reflection in a mirror, most animals display social responses (e.g., threats, bobbing, and vocalizations) toward their own image (see Anderson, 1984a, for a review). Monkeys, for example, respond to their mirror reflections with reactions usually directed toward conspecifics and, at the same time, with behaviors usually directed toward objects. The following description illustrates these various reactions as they are typically expressed by stumptail macaques confronted for 1 hour to a 30-cm round mirror hanging from a string: "Most individuals would . . . go behind the mirror, and some, while holding it with one hand, would sweep the air behind it, as if they wanted to find the animal behind. They would also sniff their reflection and touch it; they also manipulated the mirror, bit it, licked it, and tried to tear it apart, as they did with other objects" (Bertrand, 1969, p. 150). After the animals have repeatedly been exposed to the mirror image, visual exploration and social responses decline and the animals may end up avoiding mirror-image stimulation (Gallup, 1975).

The Mark Test

In contrast to other animal species, chimpanzees and orangutans confronted with their reflection in a mirror start to display "self-directed" behaviors, such as using the image to look at normally invisible parts of their body (e.g., the top of the head or the inside of the mouth: Gallup, 1970; Lethmate and Dücker, 1973). A formal test of self-recognition was developed by Gallup (1970). Four chimpanzees are exposed for 80 hours (over a 10-day period) to a full-length mirror positioned outside their cages. On the eleventh day, each chimpanzee is anesthetized and marked on a eyebrow ridge and the top half of the opposite ear with a bright red, odorless, and non-irritating dye. Then the subject is returned to its cage and allowed to recover in the absence of the mirror. Following recovery, all subjects are observed in order to determine the number of times the subject touches any marked portion of its face in the absence of the mirror. The mirror is then reintroduced as an explicit test of self-recognition. Upon seeing their marked faces in the mirror, all the chimpanzees make repeated mark-directed responses. They try to touch and inspect the marks while watching the image, and some subjects visually examine and smell their fingers after touching the marks. Moreover, on the day of testing, the chimpanzees spend four times as long viewing themselves in the mirror as they had prior to the test.

Most primate species other than chimpanzees and orangutans that have been mirror-tested do not pass the mark test (Anderson, 1984b). Even the gorilla, the remaining great ape, failed (Suarez and Gallup, 1981; Ledbetter and Basen, 1982; but see Patterson and Cohn, 1994, for convincing evidence of self-recognition in a sign-trained gorilla). Developmental studies with a large group (105) of chimpanzees have revealed that mirror self-recognition generally emerges between 4.5 years and 8 years of age, and that this capacity may decline in adulthood (Povinelli, Rulf, Landau, and Bierschwale, 1993).

The interpretation of the mark test is a controversial matter. For Gallup (e.g., 1970), mirror self-recognition implies the representation of a concept of self-awareness. For others (Kummer, cited in Dasser, 1985), the chimpanzees' behavior in front of the mirror is an expression of their ability to form "a concept of their own body" rather than "a person-concept." But alternative interpretations are available. For example, Heyes (1993c) considers that the mark-touching effect is an artifact of the anesthesia procedure. Moreover, this author challenges the view that the mark test provides convincing evidence that any primate can use a mirror as a source of information about its own body. Instead, the chimpanzee only demonstrates, for Heyes, an ability to use novel, displaced visual feedback about its physical state and behavior.

Mirror-Guided Activities in Animals

Both primate and nonprimate species are able to gain information from novel visual feedback. These abilities are tested by tasks requiring the guidance of hand movements in space.

Anderson (1986) reported that two macaques spontaneously use a mirror to direct manual searches for otherwise invisible food targets (see also Itakura, 1987). This ability was also demonstrated in two grey parrots by Pepperberg et al. (1995). A similar task proposed to two Asian elephants revealed this species' ability to locate hidden food through mirrored information (Povinelli, 1989). Note that these same elephants fail to show signs of self-recognition as assessed by the marking procedure. Chimpanzees (Menzel, Savage-Rumbaugh, and Lawson, 1985) can use televised images to guide their hands to targets that cannot be seen directly. These displaced hand movements by chimpanzees are more complicated than the direct searches performed by macaques and elephants, since they require the subjects to process inverted television images of their arms.

Another type of self-directed mirror-guided response has been described in pigeons (Epstein, Lanza, and Skinner, 1981). After being systematically trained to peck at dots placed on parts of their bodies and to look in a mirror to see a spot of light appearing on a wall and to peck at it later, the pigeons are marked with dots on their chest. The birds are equipped with a collar that prevents them from directly perceiving the dots, however. When, during the test, a mirror is placed in the experimental chamber, a pigeon will approach the mirror and rapidly move its head toward the position on its body that corresponds to the hidden dot. Even though the authors of this work claim that they have studied "self-awareness" (their words!) in the pigeon, the distinction proposed above between R0 and RAM explanations must be applied to their argument. Because of its exclusive focus on the reaction by a pigeon to its image and not, for example, to the image of another pigeon, this study does not allow one to conclude that the bird possesses or has built an image of self (RAM explanation). Furthermore, it suffers from a number of failures to replicate and has been criticized on methodological and conceptual grounds (e.g., Thompson and Contie, 1994).

Also relevant in this discussion is the use of potential alternatives to the mirror-guided reaching test (in chimpanzees: Savage-Rumbaugh, 1986; in rhesus monkeys: Rumbaugh, Richardson, Washburn, Savage-Rumbaugh, and Hopkins, 1989), such as testing the animal's control of a cursor to locate targets on a monitor (see also Figure 2.4). In such studies, there is a one-to-one correspondence between the movement of the joystick and the movement of the cursor. One of the tasks proposed to the subjects requires the tracking and hitting of a target moving randomly on the screen (Vauclair and Fagot, 1993). Another task requires the subject to maintain the cursor in contact with the moving target for durations ranging from 1 to 10 seconds (Rumbaugh et al., 1989). The tracking task requires the animal "to keep its eye on the screen and to use the cursor position as a source of novel, displaced, visual feedback on the position of its hand" (Heyes, 1993c, p. 916).

If monkeys can solve a tracking task (which, according to Heyes, involves the same kind of cognitive processing as the mirror recognition test) but fail to pass the mark test, it may be because they obtain food rewards in the tracking test but not in the mark test.

Another argument can be advanced to explain the behavioral differences between monkeys and apes (mostly the chimpanzee) when confronted with the mirror. This argument takes into account social constraints, namely that monkeys generally avoid direct eye-to-eye contact with a conspecific, which usually signals aggression. This emotional meaning, if associated with eye contact with their own reflections in the mirror, could prevent monkeys from detecting the contingencies between the features of their own bodies and those of the reflected image. The contingencies offered by mirror images and the intermodal equivalence needed to elaborate a representation of its own body might consequently be denied to the monkey. Some experimenters have found a way around the problem of direct eye contact. In a study with capuchin monkeys (Anderson and Roeder, 1989), the subjects are presented with mirrors at angles (e.g., 60°) that do not allow eye-to-eye contact. The capuchins made fewer social responses to the tilted mirrors, but they did not engage in any self-directed behaviors either. Novel attempts to investigate the necessary and sufficient conditions to pass the mark test and a critical review of the methodology and interpretation of the mirror studies can be found in Thompson and Boatright-Horowitz (1994).

Control of Subjective States

The issue of "theory of mind" (see above) refers to the way animals adapt to the mental states of others. Increasing attention to this issue must not lead us to neglect inquiries into the way animals might monitor their *own* mental states (Schull and Smith, 1992). Given the difficulty of experimentally assessing subjective states in animals, very few studies are available in this domain.

One study (Beninger, Kendall, and Vanderwolf, 1974) examined the ability of rats to respond differentially according to their own behavior. Rats are trained to press one among four levers when a buzzer sounds. This buzzer signals the availability of a food reinforcement. The correct response depends on the behavior the rat is engaging in at the time of the buzzer onset. Four frequent and spontaneous behaviors are considered: immobility, face washing, walking, and rearing. Each lever press is associated with the occurrence of one of the four behaviors. The results indicate that the rats can discriminate the four behaviors. Moreover, detail behavioral analysis discounts the possibility that rats are using cues from their locations in the experimental room or from fields of visual stimuli. "The abilities of the rats to discriminate between several of their own behaviors by responding differentially on a number of levers appear to resemble the verbal discriminations of their own behaviors made by humans" (Beninger et al., 1974, p. 90). This research thus represents one approach we might take to understand the "subjective" aspects of the behaviors of animals.

How monkeys monitor objective or subjective states of uncertainty about a discrimination problem has been experimentally studied using the joystick procedure described above (Smith, Schull, Washburn, and Shields,

1994). The subjects are two rhesus macaques, which move the cursor (by way of joystick manipulation) to one of three objects appearing on the monitor screen: a box, an *S* pattern, or a star pattern. Each object occupies a different location on the monitor. The task for the animal is to discriminate between two different densities of dots displayed in the box (dense and sparse). At first, only easily discriminated dense and sparse stimuli are presented (i.e., 2,950 pixels and 450 pixels). If the stimulus has a density of 2,950 pixels, the correct response is to contact the box with the cursor; if its density is less than 2,950 pixels, the correct response is to contact the *S* pattern. Once this basic discrimination is established, task difficulty is increased by presenting novel densities (increments of 100 pixels from the sparse density). As the task becomes more difficult, the star pattern is introduced. Touching the star clears the screen and initiates a guaranteed-win trial—which means the monitor displays only the box or the *S*. Exclusive use of the star delays the arrival of the guaranteed-win trial and, in fact, has an effect similar to that of producing incorrect responses (i.e., introduction of a delay before the next trial is presented). Thus, the optimal strategy for the animal is to attempt the primary discrimination whenever possible and to escape only the most difficult trials by using the star option.

The macaques are able to make basic discriminations after four hours of training. Both subjects also transfer the use of the star to probe trials as soon as these trials become difficult. The pattern of results can be summarized as follows: when very sparse stimuli are presented, *S* responses predominate, while box responses predominate when dense stimuli are shown. Where the two response curves cross, the discrimination is performed at chance. It is precisely in this region of maximum uncertainty that most star responses appear (see Figure 7.5).

The authors conclude that the two monkeys assess when they are at risk for error and adapt their response (choose the star) accordingly. Human subjects tested with this task basically show the same pattern of results as the monkeys. Moreover, when they choose the box and the *S*, human subjects explain their choice by referring to objective stimulus conditions (i.e., the density of dots), but they always refer to subjective mental states when they choose the star ("I just wasn't sure"). Given this similarity in results, one may wonder whether the escape (star) responses of the monkeys have the same meaning as the uncertainty-based responses made by the human subjects. It is, of course, impossible to evaluate exactly the

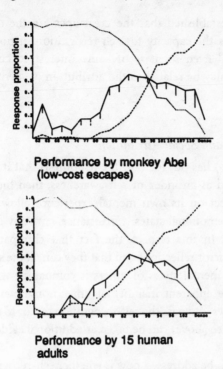

Performance by monkey Abel
(low-cost escapes)

Performance by 15 human
adults

Figure 7.5. Results of a test of uncertainty monitoring for one monkey and for 15 human subjects. The horizontal axis indicates the difficulty of the probe trial, which could be increased in 100 steps. The solid line represents the proportion of trials at a particular difficulty level on which the subject made the bailout response. The error bars show the lower 95% confidence limits. The proportion of remaining trials ending with each primary response are indicated by the dashed and dotted lines. (From Smith et al., 1994.)

subjective correlates of the monkeys' responses. Nevertheless, Schull and Smith (1992) conclude that the monkey's "bail-out responses reflect metacognitive reactions to subjective uncertainty rather than low-level reactions to (e.g.) objective stimuli or conflicted behaviors. That is, monkeys bail out when they know they do not know the solution of a trial" (p. 167).

Relationships between Mirror Recognition, Social Attribution, Imitation, and Teaching

All the studies reported so far implicitly or explicitly draw on capacities related to the concept of "self" or "other." At this point, the evidence is not definitively established that the great apes are the only nonhuman species to possess the capacity for self-recognition, if not for forming a concept of "self." Even so, it is of some interest to contemplate how self-recognition may be related to the attribution of mental states, imitation and teaching.

Self-Recognition and Social Attribution

Gallup (e.g., 1983) has put forward the hypothesis that if self-recognition can be considered as an index of self-awareness, then the capacity of an organism to reflect on its own mental experience allows it to attribute psychological or emotional states (for instance, empathy or deception) to other organisms. In this respect, the fact that chimpanzees appear to recognize their mirror reflections and that they can, at the same time, make inferences about mental states of others is compatible with Gallup's hypothesis. And the apparent inability of other nonhuman primates—for example, macaques—to pass the self-recognition test or to attribute mental states to others (see above) can be taken as additional evidence supporting Gallup's theory.

The question to be addressed now is whether self-recognition can really be taken as "an empirical marker" of theory of mind (Povinelli, 1993). The preceding section has shown that positive responding on the mark test might not be the only criterion for assessing a subject's ability to show some control of the "self." Moreover, it might be too simple to conclude that there is a causal relation between self-recognition and social attribution. In effect, as pointed out by Povinelli (1993; see also above), the distinction between perceiving and knowing has produced results which are difficult to accommodate within Gallup's framework. For example, while 3-year-old children (but not 4-year-olds) find it difficult to understand the formation of knowledge in others (e.g., Povinelli and De Blois, 1992), this form of social attribution seems to develop rather independently of, and later than, self-recognition. Self-recognition in human infants is present by approximately 18–24 months of age (Anderson, 1984b; Brooks-Gunn and Lewis, 1984). The first recognition of contingencies between the self's movements and those of the mirror appears earlier (8–12 months: Lewis and Brooks-Gunn, 1979), and infants as young as 3 months of age show differential cardiac and behavioral responses to their own mirror images and same-sex infant peers (Field, 1979). In brief, it can be stated relatively safely that self-recognition per se is not sufficient for social attribution.

An additional example of a dissociation between self-recognition and theory of mind in humans can be found in autistic children. Although they engage in self-directed behaviors in front of a mirror (Neuman and Hill, 1978) and are able to display, as do chimpanzees, rudimentary forms of social attribution, such as "perspective taking" (Leslie and Firth, 1988), these children are apparently unable to attribute certain complex forms of mental states, such as an understanding of false beliefs (Leslie, 1987).

In summary, the available evidence suggests that the presence of self-recognition in humans is not a sufficient indicator of complex social attribution. Self-recognition should rather be considered as an "empirical marker of the beginning of a complex developmental sequence involving multiple feedbacks between self- and social attribution" (Povinelli, 1993, p. 503). In other terms, as suggested by the cases of autistic children and chimpanzees, self-recognition is correlated with some but not necessarily all forms of social attribution.

Imitation and Teaching

An important feature of self-recognition in humans is its dependence, in the course of the development, on contingent play, a sort of imitative behavior that includes facial movements and peek-a-boo games (Lewis and Brooks-Gunn, 1979). These behaviors emerge around 1 year of age and correspond to Piaget's stage 5 of the sensorimotor period (see above and Chapter 3). Lewis and Brooks-Gunn have shown that the cognitive measures most strongly correlated with self-recognition are contingent play (expressed, for example, in making faces or playing peek-a-boo) and imitation. And it is also likely that when the child realizes that the image in a mirror is moving when he is moving, he might also understand causality in external objects (Parker, 1991). Studies with dyads of unfamiliar 19-month-old children (Asendorpf and Baudonnière, 1993) show a positive correlation between mirror self-recognition and synchronic imitation (i.e., the two children play at the same time with the same type of objects in a more or less similar way).

I have already noted that monkeys rarely engage in imitative behaviors—for example, during tool use (Visalberghi and Fragaszy, 1990). One reason for this absence of imitation in animals, in particular in monkeys, might be a lack of understanding of causality: the relations that link objects, actions, and their outcome (Visalberghi, 1992). We have seen that novel behaviors are most often acquired by animals by observational learning, that is, learning by observing a competent individual performing a given behavioral sequence. On some occasions, animals can perform imitation or copying but, contrary to what is commonly observed in humans, systematic monitoring or teaching has rarely been described as a means of acquiring a novel behavior (see also Chapter 6). It is important to note that although teaching usually refers to the deliberate transmission of information to others, some authors (e.g., Nishida, 1987) also consider *dissuasive* behaviors directed toward the young as cases of teaching. For example, Japanese macaque adults have been seen to move infants away from novel objects. I have similarly observed adult baboons preventing infants from getting too close to the electrical fence of their outdoor enclosure.

Pedagogy in the Chimpanzee?

A very accomplished case of monitoring by competent adults during the acquisition of complex behaviors in immatures has been reported for wild chimpanzees (Boesch, 1991, 1993a). This example concerns the development of a tool-use behavior, pounding nuts with hammer stones or clubs, by chimpanzees in the Ivory Coast (see Chapter 4 for a description of this behavior). Field observations indicate a rather unusual involvement of the mother in the acquisition of nut-cracking skills by the young (it must be recalled that almost 10 years are needed to fully master the technique: Boesch and Boesch, 1990). Boesch (1991, 1993b) describes various stimulating and facilitating acts on the part of the mothers in this group of chimpanzees that affected the development of the nut-cracking technique by their offspring. "Stimulation" occurs when the mother leaves hammers or nuts near the anvil. This behavior has never been observed in chimpanzees without infants. "Facilitating actions" occur when mothers provide good hammers or when they provision their infants with nuts. Although these behaviors are relatively frequent, behaviors more similar to teaching have been observed on only two occasions: when a 6-year-old male attempted to crack a nut, his mother corrected the position of the nut on the anvil; when a 5-year-old female tried unsuccessfully to open nuts, her mother approached after a while, seized the hammer, and "in a very deliberate manner, slowly rotated the hammer into the best position with which to pound the nut effectively" (Boesch, 1991, p. 532). The juvenile eventually adopted the grip demonstrated by the mother.

This latter example notwithstanding, the general picture of animal learning excludes direct and systematic interventions by a competent adult such as a mother (see Heyes, 1994, for a review of social learning in animals). Although a systematic intervention on the part on the mother can be excluded, the literature gives an incomplete picture of the roles of individual discovery and social mechanisms, including imitation and teaching, in the development of tool use in chimpanzees. From the few reports available on apes' acquisition of tool use, it appears that nothing close to the systematic shaping of tool use that occurs in humans is found in apes. McGrew (1977) claims that wild chimpanzees "are capable of acquiring a complex tool-use technique with only limited exposure and practice" (p. 284). In fact, the mother-offspring relationship, with its long period of dependency, has the effect of repeatedly exposing the infant to certain important environmental stimuli and events, thus increasing learning opportunities.

The effects of viewing a competent tool user on subsequent performance have been studied experimentally (Tomasello et al., 1987). Young chimpanzees who learn the tool-using task do not copy the demonstrator's precise behavioral sequences, but they are nevertheless quicker to learn than subjects not exposed to the demonstrator. A direct comparison of juvenile chimpanzees and 2-year-old human children confirms that the apes can benefit from watching a skilled demonstrator but do not imitate the most efficient form of tool use, whereas the human children do imitate (Nagell, Olguin, and Tomasello, 1993). Further and more systematic comparisons of imitative learning between three different groups—mother-reared chimpanzees, chimpanzees living in a human-like sociocultural environment, and human children—reveal important differences between the two ape groups. The mother-reared chimpanzees appear to be poorer imitators (immediate imitation) than "enculturated" chimpanzees, while the latter group does not differ from the children (Tomasello, Savage-Rumbaugh, and Kruger, 1993).

Attributional Capacities and Teaching

In the context of the present discussion, an absence of teaching might result from the inability of adults to attribute mental states to others and particularly to their offspring (Premack, 1985). It must be stressed that the chimpanzee might represent an exception in this respect, given that this species seems to have access to self-recognition, as well as to rudimentary forms of social attribution and teaching. The pattern most likely to occur in animals, and in particular in monkeys for which direct experiments have been carried out, is a limited control over their internal states, including an inability for self-recognition and no attribution of mental states. If this scenario is correct, it is not surprising that monkeys should also lack an equivalent of "pedagogy" in the sense we use this term with humans. "To teach, one must recognize a difference between one's own knowledge and someone else's knowledge and then take explicit steps to redress this imbalance. Without attribution, instruction cannot even begin, because those with knowledge do not realize that the information possessed by others can be quite different from their own" (Cheney and Seyfarth, 1990a, p. 306).

Summary and Current Debate

To summarize and conclude this chapter, I would like to stress that the topic of theory of mind has recently become a very active and exciting issue in the field of comparative psychology. Research efforts in this direction are largely indebted to the pioneering work of Gallup (e.g., 1970) and to the new fields of investigation initiated by Premack and colleagues (e.g., Premack and Woodruff, 1978), whose proposal of theory of mind has since become popular among developmental psychologists (e.g., Wellman, 1990).

The difficulty of research in this area has already been underlined, as has the caution that must be taken to avoid overinterpreting or misinterpreting the findings of these studies (e.g., Heyes, 1993b). In this line of work more than in any other, the researchers must never forget the principle of parsimony (or "Morgan's canon": see Chapter 1).

Currently there is some debate over how to show evidence of self-recognition (if not of self-awareness) in animals. The discussion in this chapter of the "mark test" outlined the controversies concerning this issue.

Still other problems arise. First, not all chimpanzees tested exhibit the typical pattern of self-recognition (Swartz and Evans, 1991; Povinelli, 1993). This lack of generality leads one to question the validity of Gallup's test as a general measure of the capacity of the chimpanzee for self-knowledge. We must search for alternatives to the mark test, for two reasons.

The first is logical: failing the mark test does not automatically imply that the animal lacks the capacity the test is supposed to measure. The second is empirical: it is conceivable that the use of the mark test, although it reflects an obvious ability to use self-generated mirror images, might actually hinder other abilities, antecedents of self-recognition.

A book edited by Parker, Mitchell, and Boccia (1994) recommends several behavioral approaches to replace the mark test in the comparative study of self-awareness. Among the suggestions are self-directed behaviors other than those directed to the face, embarrassment or coy reactions to the mirror image, contingent play with body movements and facial expressions, and shadow recognition. "These measures have the added advantage that they correlate with such other traditional cognitive and affective developmental measure as stages of object permanence, causality, imitation, and symbolic play, thereby allowing various stages of self-awareness to be placed in a larger cognitive/affective framework. In addition, as spontaneous behaviors, they have the advantage of being observable both inside and outside the laboratory setting" (Parker, Mitchell, and Boccia, 1994, p. 15).

Moreover, as the section on mirror-guided activities in animals has demonstrated, pigeons, elephants, and monkeys can be trained to use the image in a mirror to search for hidden food. More indirect uses of images (control of the displacement of a cursor on a monitor) to control hand movements have also been obtained with monkeys (e.g., with baboons: Vauclair and Fagot, 1993). It may be that video tasks can be used successfully to investigate precursors of mirror self-recognition in monkeys. In effect, this task (see Chapter 2 for a description) may be employed to test the subject's capacity for differentiating images on the screen, by recognizing the contingency between its movement of the joystick and the movement of the image on the screen (note the similarity with mirror-guided reaching). This would imply an ability to differentiate between self-generated, visually displaced contingent movements and other-generated movements on the computer screen. Jorgensen (1994) has successfully tested capuchin monkeys and chimpanzees in variants of the joystick task, by disrupting the contingency between the movement of the joystick and the cursor or by adding a competing image that the subjects must differentiate from the image they control. These findings suggest that the subjects can differentiate between self-generated and other-generated movements, and they can recognize the contingencies between self-generated movements and displaced visual movements. These two abilities, as they are expressed through this novel technique, are essential requirements for self-recognition (see Heyes, 1993c), but they are unlikely to be sufficient for the acquisition of self-recognition.

In sum, the growing number of techniques for comparing the way monkeys, chimpanzees, and human children distinguish between "self" and "other" should not only enhance our understanding of self-recognition and of its antecedents. It should also stress the importance of the non-human primate as a model, according to the terms used by Povinelli (1993), for "reconstructing the evolution of metacognition."

8

An Agenda for Comparative
Cognitive Studies

Over the past decades many advances have been made, in both theory and experimental research, in our understanding of animal cognition. We are now poised to ask more general questions about the field of comparative cognitive studies and its place among biological and psychological investigations of animal behavior. In this book I have focused on the experimental approach to cognitive studies, but in this final chapter I will address other major fields of inquiry, such as cognitive ethology (Griffin, 1978). I will also consider alternative approaches to the study of animal cognition, such as the ecological approach based on comparisons between species (e.g., Kamil, 1978a). I conclude this summary by proposing a possible classification of representational systems constructed by animals.

Cognitive Ethology: Mental Representations or
Mental Experiences?

The concept of consciousness was banned from experimental psychology when behaviorism imposed itself at the beginning of this century (see Chapter 1). This position is still maintained today: "Just as the modern rationale for using human cognitive terms is not based upon arguments that appeal to consciousness or to introspective reports, the rationale for the study of cognitive processes in animals requires no reference to animal consciousness. Both in human and animal cognition it is assumed that the normal state of affairs is unconscious activity and thought" (Terrace, 1984, p. 8).

Notwithstanding this recommendation, a field known as "cognitive ethology" emerged after the publication of *The Question of Animal Awareness* by Donald Griffin (1976). This book and other publications (e.g., Griffin, 1978, 1984; Ristau, 1991b) aimed to rekindle scientific interest in the "conscious mental experiences of animals." On the basis of ethological studies (for example, the work on the "language" of bees by Von Frisch; see Chapter 6), as well as the chimpanzee Washoe's use of a gestures for communication (Gardner and Gardner, 1969; see Chapter 6), Griffin concluded that the time is ripe to consider phenomena related to consciousness in animals. In effect, he proposed that the social domain, and in particular the field of intraspecific and interspecific communication, constitutes a "window on animal minds." Moreover, Griffin states that the reluctance expressed by contemporary scientists to investigate the subjective mental experiences of animals could "result, in part, from unrecognized vestiges of behaviorism that inhibit inquiry" (Griffin, 1991, p. 4). Although Griffin is undoubtedly right that animal behavior is guided by complex thinking processes, he does not propose specific methods to study in animals what he calls the "ambassadors of thinking." Moreover, and surprisingly, Griffin's theory does not refer to cognitive concepts such as representation. In more global terms, his endeavor does not rely much on studies of comparative psychology or even on psychology in general.

Several questions have been raised about Griffin's project (for details, see the commentaries to Griffin's paper of 1978). One area of concern is what is understood by consciousness in cognitive ethology. Griffin himself seems to use this concept in a rather broad sense, since consciousness is synonymous with mental states, thinking, or mind. It appears that this vagueness in terminology might hide a semantic ambiguity between mental experience and mental representation. Ullman (1978) has argued that Griffin commits two fallacies. The first amounts to identifying self-awareness with self-recognition. As suggested in Chapter 7, several interpretations may account for the capacity of some species (orangutans, gorillas, and chimpanzees) to recognize their reflections in a mirror. A parsimonious interpretation claims that this behavior reflects at best an ability of the animal to represent its own body. Consequently, the evidence that some animals may have human-like mental experiences still awaits demonstration.

A second possible fallacy in Griffin's position is related to the identification of the capacity for planning future activities with conscious intentions

(Ullman, 1978). Anticipatory activities are the basis of most adaptive behaviors, and they in no way imply the intervention of intentional or conscious processes. As Chapters 6 and 7 have shown, intentionality is reserved for certain special forms of social exchanges for which a number of criteria must be met, such as the sender of a message controlling the content of the message and of its consequences on the receiver. Anticipatory capacities do not require that such a complex construction take place. In effect, if cognition primarily consists of the detection of regularities among events in the environment, it is likely that an organism will quickly detect and anticipate these regularities. For example, Von Frisch (1967) has observed that bees trained to fly toward a food source whose position changes from trial to trial display some sorts of anticipatory behaviors: some bees, as if they understood that the experimenter always places food beyond the last visited location, anticipate these displacements and immediately fly to a new location where food might be found.

Several other criticisms have been raised concerning cognitive ethology (e.g., Mason, 1976; Wasserman, 1993). In addition, critics have put forward suggestions for integrating this approach within the methodology and theories of human and animal cognitive psychology (Yoerg and Kamil, 1991).

Types of Representations and Their Substitution Systems

In this book I have stressed the existence of different forms of representations in all animal species studied, from invertebrates to vertebrates. Examples concerning spatial coding were provided for insects in Chapter 4, and other examples could be provided for invertebrates (as in molluscs: Young, 1991) or for lower vertebrates (as in reptiles: Burghardt, 1977). I have also discussed numerous examples for birds, rodents, small and large mammals, and of course primates. If a capacity for representation is common to various zoological groups, it becomes tempting to suggest a classification of the representations as a function of their complexity. This task is difficult, however, mostly because research efforts have not been evenly distributed among zoological groups. Much emphasis has been put on nonhuman primates, less on other species.

To posit the existence of representational capacities in animal species means that one conceives of organisms as active devices that filter, process, and transform different information coming from the environment. But to say that organisms elaborate representations is no more informative than stating that organisms learn when they have to adapt to the constraints of their environment. Further characterization of the representational processes is necessary if one wishes to give a heuristic value to the concept of representation—namely, that it serves to enlighten our understanding of how animals face the demands of changing environmental conditions.

Two aspects of representations may be analyzed in this way: the first concerns the substitution systems that make representations possible, and the second concerns the type of mental operations that representations carry out.

Substitutes as Representations

The representation of an object or event is not an exact replica of the original but a somewhat simplified, partial, or abstract "copy." The linguistic concept of *substitute* helps us differentiate the substitution mechanisms of representation (see also Chapter 6). The most primitive substitutes are certainly *perceptual indices,* which are connected in time and space to the object to which they refer. Indices are either a part of an object, a temporal antecedent, or a causal result (for example, smoke is an index of fire). A small part of an object (for example, the nipple of a feeding bottle) that serves to identify the object as a whole is a perceptual index for that object. There are also *internal substitutes,* which are real, internalized representations of an object. The prototype for an internal substitute is a mental image that permits the subject to evoke the object in its absence. Two groups of external substitutes can be distinguished, *external analogical substitutes* and *external arbitrary substitutes.* Analogical substitutes share a physical or geometrical resemblance with the referred object (this is, for example, the case for miniature models or maps). Arbitrary, or conventional, substitutes have no similarity to the object. Linguistic signs and systems that derive from language (arithmetic notations, for example) are arbitrary substitutes.

Even though this semiotic classification of substitutes (see Table 8.1) encompasses a relatively limited number of elements, it should make it possible to outline a framework for evaluating the modes of symbolization found in animals. For example, it is a distinctive feature of the world of primates that they are quick to learn to use objects with arbitrary features to represent other objects; this ability is exemplified by language-trained chimpanzees, which use plastic tokens or keyboard keys to refer to objects, actions, qualifiers, and so on (see Chapter 6).

Table 8.1. Substitutes of representations and a classification of their types.

Substitutes

1. Perceptual indices
2. Internal substitutes (e.g., a mental image)
3. External substitutes
3a. External analogical substitutes (e.g., a map)
3b. External arbitrary substitutes (e.g., a linguistic sign)

Types

1. Temporal and spatial couplings
2. Comparisons (e.g., judgments of similarity and difference)
3. Relations (e.g., categorization and ordering)
4. Equivalents of logical operators (e.g., inference, transitivity)

In contrast, the ethological literature is rich with examples of the use of objects in courtship contexts (for example, the use of nest material by great-crested grebes reported in Manning and Dawkins, 1992). Using objects as offerings and appeasement has been described in insects and birds (see references in Eibl-Eibesfeldt, 1970), and it would certainly be wrong to limit the utilization of external substitutes to primates. Yet it is unlikely that insects or birds could attain the level of sophistication displayed by primates in the use of external substitutes.

An experiment performed long ago with chimpanzees (Wolfe, 1936) illustrates this point. The chimpanzees were first trained to manipulate a tool to reach for food. Then, the food reward was replaced by a token that can be later exchanged for food: the chimpanzees continued to use their tools to gather tokens. These tokens functioned as substitutes, or "abstract tools," that were piled up and even stolen by some chimpanzees in order to trade them later on for food.

The description of the different substitutes must be complemented by a consideration of the levels of processing at which the representations are elaborated.

Levels of Representational Processing

The representations constructed and used by an active organism are processed at a variety of levels (see Table 8.1). Several of these levels have already been detected (Thomas, 1980, offers, from the viewpoint of learning theories, a hierarchy of intellective abilities that fits in with the framework presented here). The most elementary and widely distributed level concerns *temporal and spatial couplings.* These couplings are manifested in associations between stimuli when two or more events, objects, or locations are perceived at the same time or in the same place. Most instances of conditioning and instrumental learning fall into this category. For example, in Pavlovian conditioning, learning occurs (that is, a conditioned response, such as salivation, is produced) because a conditioned stimulus (such as a tone) is paired with (is presented soon before) a biologically significant unconditioned stimulus (food). Similarly, in instrumental or Skinnerian conditioning, learning takes place because a given action (a lever press by a rat, perhaps) is consistently followed by a given consequence (a reinforcer, such as a piece of food).

The next level concerns the act of comparing objects or situations: comparisons constitute the basis of the *relations of similarity and difference.* An example is the discriminative and matching tasks presented in Chapters 2 and 5. Another level concerns the use of different types of relations, such as *categorization* and *ordering* (examples of which are described in Chapters 2 and 3). Moreover, equivalents of logical operations can be applied to the relations between animate and inanimate objects. *Inferential behaviors* bearing on spatial relations (e.g., the cognitive maps described in Chapter 4), the physical properties of objects (see Chapter 3), or conspecifics (e.g., making representations of social hierarchies, see Chapter 5) are examples of logical operations.

This classification, though rudimentary, may be sufficient for organizing our ideas about the mastery of the representational systems in different species and different contexts. For example, in some social contexts, different types of processing might be at work as individuals of different species represent their position in the group hierarchy. To take one example here, the behavior of a junco in its social group may thus be fruitfully compared with that of some primates, such as the vervet monkey or the macaque (see Chapter 4), in their social groups.

The social hierarchy of the junco has been studied in an experiment making the group compete for food (Wiley and Hartnett, 1980). A junco of intermediate rank systematically chases a lower-ranking junco from the food tray. However, this same middle-ranking bird, when *two* birds that it dominates are at the food tray, does not show any particular tendency to chase away the lowest-ranking bird. In other words, although the junco reacts differentially toward high- and low-ranking birds, it does not seem to take into consideration the social distance that separates the birds in the hierarchy. It is as if the junco recognizes all conspecifics as part of a two-class hierarchy: those which are "above" and those which are "below" a given individual. In this respect, the representation that can be inferred in the junco appears to be somewhat restricted in comparison with that observed in vervet monkeys by Cheney and Seyfarth or in macaques by Dasser (see Chapter 5). These studies with nonhuman primates have revealed that a given individual has quite a good knowledge of the relations other individuals entertain with each other in affiliative or hierarchical structures.

The representations that emerge from experimental studies of cognitive processes in animals are described in general terms, such as categorization or cognitive mapping. These descriptions express the fact that a given individual is likely to structure its perception of the social and physical environment that surrounds it. The existence of representations is nevertheless insufficient to predict how the representations will affect real behavior. At this point, it is necessary to consider that the representations will be reflected in rules of actions that can take various configurations.

The Distinction between a Representation and a Rule of Action

At first glance, the translation of a representation into a rule of action, or response, appears to depend upon constraints related to variability in the environment, training history, or species characteristics, such as feeding habits. Some examples will serve to outline the distinction between a rule of action and a representation. These examples come from the field of spatial cognition. The first example concerns the behavior of rats placed in the radial maze (a maze made of eight arms radiating from a central platform; see Chapter 4). Food is left at the end of each arm, and the task for the animal is to visit the eight arms and to collect all the available food. Olton and Samuelson (1976) have shown that, after a few trials, the rats become very efficient in collecting the food, since they rarely visit the same arm more than once per trial. The apparent response rule that is reflected in the rat's ability to avoid visiting the same arm twice is the "win/shift" strategy. It must be kept in mind that rats can successfully search a radial maze even if food is not left at the end of arms, an ability that might have resulted from an evolutionary adaptation for efficient foraging and increasing environmental familiarity in the absence of food (Timberlake and White, 1990).

The "win/shift" rule has also been described for some nectarivorous birds that avoid revisiting flowers they have recently emptied (Kamil, 1978b). In fact, these birds have developed a memory for emptied flowers.

"Since in many cases flowers produce nectar at predictable rates, nectar feeding birds may be able to fine-tune the temporal patterning of their flower visitations. Since immediate returns to already visited flowers never produce reward, while visits to unvisited flowers often do, nectarivorous birds may find it easier to learn to shift among different positions than to learn to return to the same position" (Kamil, 1984, p. 536).

By contrast, in other contexts, and for those animal species in which spatial memory is at work (see, for example, the case of the marsh tits able to store and then recover hundred of seeds: Shettleworth and Krebs, 1982), the response rule may *not* be the "win/shift" strategy. The marsh tit tends to return to the last-visited site (this is known as the "recency effect"; see Chapter 2), because seeds searched for in recently established caches are likely to be found most reliably. This strategy is thus adapted to diminish the effects of stealing by other animals and of natural degradation. Among nonhuman primates, marmosets appear to prefer using "win/stay" strategies in simulated foraging tasks (MacDonald, Pang, and Gibeault, 1994).

The value of distinguishing between different expressions of representational systems in animal cognition is not recognized equally by all scholars. Two of the positions currently at variance with the mainstream of comparative psychology are illustrated by the work of Dickinson and Macphail.

Dickinson (1980) contends that all animal species build representations and that representations concern causal regularities between events in the environment that have an adaptative significance. The ability to make representations can be seen in all forms of learning, both simple and complex. Dickinson reinterprets the conditioning experiments of others (e.g., Holland and Straub, 1979) and asserts that internal representations are stored by the animal in a propositional way. For example, the representation of a given event (such as the sound of a bell) is a precursor of a second event (the delivery of food that causes illness in the animal). The rat is assumed to be able to integrate separate elements of information into the whole "proposition": "Sound causes illness." This conception, though it postulates a general capacity for representation, does not consider different levels in the representational process. Nevertheless, this conception accepts the cognitive position that the animal stores the information in a way that is flexible enough to be combined with other relevant information about the environment.

A different perspective is offered by the theory put forward by Macphail

(1982, 1987) with the aim of detecting the most general mechanisms of intelligence in animals. These mechanisms are described without reference to concepts used in cognitive studies (such as the notion of representation). Rather, they refer to the formation of associations between stimuli or events in the environment and to the responses produced by the organism: "Association formation is a process which evolved to detect causal links between events . . . [it] is a necessary prerequisite of useful prediction in all environments" (Macphail, 1987, p. 655). Moreover, these mechanisms are of general applicability and do not seem to be dependent on any specific ecological niche.

More importantly, Macphail hypothesizes that there is no difference in these mechanisms among different species, at least as far as vertebrates are concerned. If association formation is the "central process of intelligence," however, humans are distinguished from other vertebrates by their possession of an additional system named the "species-specific language-acquisition device".

Several problems are raised by the "null hypothesis" defended by Macphail—that is, that there are no differences among species in how associations are formed (see the commentaries to Macphail, 1987 and 1990, and Pearce, 1987). For example, this general conception leaves little room for the occurrence of any forms of learning other than association (for example, observational learning; see Chapter 7). Neither does it take into account the role of "biological constraints" on learning (see below). Another objection to this position is that cognitive concepts, such as those presented in this book, may be employed to differentiate mechanisms used by animals when they deal with animate and inanimate objects, especially for species with markedly different ecological niches. The effect of the niche as well as other kinds of constraints (for example, foraging needs) on the organization of cognitive processes is examined in the following section. In brief, the view that language is the distinguishing characteristic of humans, and that language is wholly absent in nonhumans, leads Macphail to deny numerous psychological phenomena observed in animals and to prefer using the more general but still undefined concept of intelligence (Macphail, 1993).

The Generalist versus Ecological Approach to Animal Cognition

Renewed interest in studies of comparative cognition has arisen concomitantly with the new wave of cognitive sciences (Gardner, 1985). This development has resulted in a growing attention to the study of cognition in nonhuman species. The most important debate currently at issue in cognitive studies is not so much about the appropriateness of the different theories or models of cognitive processes as it is about which behaviors can best tap the reality of this cognition in nonhuman species. The two sides of this debate are usually described as the generalist view and the ecological view (Riley and Langley, 1993). (See Table 8.2.)

As I explained in the first chapter, Darwin's theory of evolution spawned the development of comparative psychology and learning theories. These disciplines were later influenced by human psychology, from which comparative psychology borrowed its theoretical concepts. Ethology, defined as the investigation (mostly through observational methods) of spontaneous and adapted behaviors of animals, also has its roots in the Darwinian concepts of selection and adaptation. According to several authors (Thorpe, 1979; Hinde, 1982), ethology also emerged as a scientific discipline from field studies made by earlier naturalists and zoologists, such as Fabre, Heinroth, Craig, or Von Uexküll (references in Eibl-Eibesfeldt, 1970). A strong ethology movement developed in Europe with the leading figures of Lorenz and Tinbergen, who emphasized the particularity of their field in comparison to experimental animal psychology. For example, according to Tinbergen (1963), animal psychology is concentrated "on a few phenomena observed in a handful of species which were kept in impoverished environments, to formulate theories claimed to be general, and to proceed deductively by testing these theories experimentally"

Table 8.2. The generalist versus the ecological programs for the study of animal cognition (adapted from Riley & Langley, 1993, & Shettleworth, 1993).

	Generalist program	Ecological program
Goal	Investigate the generality and continuity of cognitive processes across species	Understand cognition as a biological phenomenon
Issues and methods	Study nonverbal thought through species comparisons (mostly in laboratory settings)	Search for cognitive processes in natural settings
	Evolution of human cognition	Compare close relatives in divergent niches and distant relatives in convergent niches

(p. 411). As the preceding chapters amply demonstrate, the boundaries between experimental animal psychology and ethology are crumbling. Nevertheless, there still exist discrepancies between the two disciplines.

The Ecological Program

The ecological program, also called the "synthetic approach" (by Kamil, 1978a), is focused mainly on the question of problem solving by animals *in nature*. For Shettleworth (1993), this program entails several ecologically relevant behaviors, such as spatial orientation, individual recognition, learning of communicative signals (e.g., song perception and learning by birds). A prototype of this program is illustrated by the studies on memory in food-caching birds (see Chapter 4). A second aim of the ecological approach is to understand the biological framework in which cognitive processes have evolved. Third, these studies stress the evolutionary relationships between species and tend to compare either closely related species confronted with different cognitive problems or to test unrelated species faced with similar cognitive problems. In brief, what this program aims to do is to describe cognitive mechanisms that underlie adaptative behaviors.

The problems posed by ethologists and behavioral ecologists are not new in the study of learning and cognition in animals. For example, the importance of biologically significant factors in the formation of learning has already been established through the concept of "biological constraints" (e.g., Hinde and Stevenson-Hinde, 1973). These constraints make learning dependent upon a species' abilities and the selective pressures that have favored, in the course of evolution, particular forms of associations between stimuli in the environment. As an example, the honey bee can learn to associate food with a particular stimulus object (an odor) in one trial. But the bee will need several trials to learn the association if the stimulus object is distinguished by color, and more than 20 trials are needed before the bee can make an association with form. In other words, a given species may have some predispositions to learn in specific contexts whereas, because of the biological significance of the stimuli encountered (or their absence), the same species may be less apt to learn in other contexts. These predispositions, along with various degrees of preparedness (Seligman, 1970) of the organism to emit one response and not the other, have challenged the universality of the laws of learning.

The Generalist Program

The present book has largely depicted the generalist approach to animal cognition. This program is generalist because it is concerned with the question whether distantly related species structure information about the environment in similar ways. Comparisons are "made to establish the generality of cognitive traits among species" (Riley and Langley, 1993, p. 185). The advocates of this approach argue that comparisons help to define the scope and the limitations of general cognitive processes. Furthermore, conducting comparative investigations with human and nonhuman species permits us, on theoretical grounds, "to restrict the range of plausible explanations [for cognition], especially to exclude those involving language" (Church, 1993). By extension, comparative research into cognitive processes in nonverbal species may provide clues to the evolution of human language (Terrace, 1993). Numerous examples can be found in the preceding chapters to illustrate the usefulness of these comparisons: the presence of serial position effects in memory processes in birds and mammals (Chapter 2), categorization and concepts (Chapters 2–7), semanticity (Chapters 5–7), inferential and deductive behaviors (Chapters 3 and 4).

The Future of the Debate: Opposition or Reconciliation?

It would be odd and inevitably counterproductive on scientific grounds (unless one wishes to encourage what Gray, 1973, called "a saga of twins reared apart") to approve of the relatively young field of comparative animal cognition engaging in a divisive debate about which approach is best suited for understanding cognitive processes in animals. An open-minded and opportunistic scientist would immediately answer that there is room for both programs and that the respective approaches can easily be viewed as complementary and not opposed. Each program has specific strengths and weaknesses.

Two main concerns may be raised with regard to the generalist position. The first is related to the lack of explanations to accompany demonstrations of generality. For example, the finding that rodents share with humans some basic representations of time intervals (see Chapter 4) should also provide information as to whether this similarity can be accounted for by similarities in their neural systems. An extension of this issue would be to determine if the similarity is purely analogical (that is, independently evolved) or homologous (in which case the trait would have been present in the common ancestor of the two species). The need to uncover the origin of a given trait calls for an explicit choice of which species to compare. In studies of animal cognition, and within the generalist current, such decisions are usually not made, and there is a tendency to limit the number of species that are studied to a handful, including birds (the pigeon), rodents (the rat), and a limited range of nonhuman primates (essentially apes, macaques, baboons, and cebus monkeys).

Several criticisms have also been addressed to the ecological approach. For example, this approach should concentrate more on closely related species that fill divergent niches and distant relatives that fill convergent niches, an important strategy if its goal is to demonstrate how cognitive processes constitute specialized adaptations to environmental demands. Other reservations about the ecological approach concern its interpretation of cognitive traits as adaptive specializations when they are not. For example, the ability of the Clark's nutcracker to cache and recover thousands of pine seeds (see Chapter 4) might not necessarily prove that its spatial memory "is a specialized adaptation that has evolved as a result of being selected for its functional advantage in recovering stored food . . . [but rather that it is] a result of selection for the ability to remember the locations of guideposts as it flew from the territory in which it lived to the territory in which it foraged" (Riley and Langley, 1993, p. 188). Thus, if spatial memory is dependent upon navigational abilities, then the cognitive ability to cache and recover food may have evolved prior to the existence of ecological conditions in which food storing has become adaptive.

A final criticism of the ecological psychology of animal cognition has to do with the fact that not all cognitive mechanisms may serve a survival function in nature or, alternatively, that cognitive traits may not be observable in nature and do not have a readily identifiable function (Riley and Langley, 1993). The above considerations pertain directly to one of the main projects of the ecological program, which is to contribute to our understanding of the evolution of the animal mind, including the human mind. The evolution of certain features of animal minds might not be the product of adaptive biological functions alone. For example, it has been hypothesized that human language could be more than the exclusive by-product of biological factors and their evolution. According to some authors (e.g., Vygotsky, 1962), language has resulted both from biological changes and social transformations. In this scenario, novel forms of representations could be conceived as biologically adapted and specialized acquisitions. But language, considered as a system that simultaneously combines representational and communicational components (see Chapter 6 and Vauclair, 1990b), can hardly be regarded as the by-product of a purely biological transformation.

The two approaches ask different questions about animal cognition and are thus complementary and not in opposition to each other. Proponents of each program should keep in mind the following statement concerning the future of comparative psychology:

> Comparative psychology has . . . lost touch with its roots in ecology and evolutionary theory. Behavioral biology has lost interest in questions of ontogeny and behavioral mechanism. Comparative psychologists do not need to reinterpret their past to justify their present or future . . . Both the search for general principles and attention to species-typical behavioral adaptations are valid approaches; either can lead to valuable insight. Availability of alternative perspectives can only enrich our discipline. (Galef, 1987, p. 260)

Animals as Problem Solvers and Decision Makers

Both generalist (Rumbaugh and Pate, 1984) and ecologically oriented (Kamil, 1984) investigators of animal cognition agree that animals in their environment can be described as problem solvers and decision makers.* Their common ground is the cognitive dimension of behaviors, which is expressed through the intervention of internalized representations of the environment by a given organism. In effect, the animal has to solve problems in almost all of its daily activities, from negotiating social exchanges to manipulating inanimate objects. The decisional aspect of animal cognition is thus similarly present during foraging as well as during social activities, as when an animal must quickly evaluate different elements in order to act appropriately (for example, flight or fight). In laboratory settings, these decisions are particularly evident in deceptive behaviors and in other forms of dissimulation.

The view that animals are active problem solvers and decision makers reinforces the hypothesis of a continuity in psychological functions between animal species, as well as between animals and humans. As stated by Walker (1983), there remain sufficient similarities between human and animal brain functions to allow comparisons if one assumes that "animal brains are devices for selecting and organizing perceived information, and that the neural systems which accomplish perception and memory exhibit evolutionary continuity" (Walker, 1983, p. 380). It appears that these global functions are performed by the animal in ways that are basically similar to human performance, that is, through the construction and use of representations of various degrees of schematization and abstraction (see also Premack, 1985).

In addition to stressing the similarities among animal species and between animals and humans, one of the tasks of comparative psychology is to point out any dissimilarities between different species. As Chapter 6 has suggested, a discontinuity might appear in the mastery of some of the codes of communication, at least between animals and humans. This discontinuity can concern communicative systems as sophisticated as linguistic signs, and also the control of intentional behavior and the development of a sense of self-awareness (see Quiatt and Reynolds, 1993, for other views on the continuity-discontinuity debate).

To return to the idea of continuity defended by Darwin (see Chapter 1), the data presented so far indicate a partial continuity among animal species as well as between animals and humans. The most obvious elements of continuity concern the presence of representational systems in certain domains, including space and memory. Some elements of discontinuity are nevertheless obvious in the domain of communicative systems, and notably with the presence of language in humans and not in other species. But the issue of language is far from resolved. First, new experimental findings may modify our views, even though the experimental evidence available so far is in favor of the conclusion that language is unique to humans. Second, the form and the content of the dissimilarity is greatly influenced by the behaviors we select to assess the difference.

The interpretation of language as a uniquely human trait may be revised if the impetus is put not on apes' abilities to *produce* equivalents of words but rather on their competence to *comprehend* language. For example, the language comprehension skills of the pygmy chimpanzee Kanzi (at age 8), whose accomplishments were related in Chapter 6, has been systematically compared to those of a 2-year-old child (Savage-Rumbaugh, Murphy, Sevcik, Brakke, Williams, and Rumbaugh, 1993). Both subjects have been exposed to spoken English and to lexigrams (see Chapter 6) from infancy. The two subjects are presented with the same sets of 660 novel sentences. Examples of sentences presented in English are "Put the milk in the water" or "Go get the raisins that are in the refrigerator." It emerges from this comparison that the subjects can understand these novel requests with comparable accuracy. Their respective comprehension of English shows that they have mastered word recursion as well as the ability to reverse word order. From these data the authors infer that the potential for language comprehension probably preceded the appearance of speech in human evolution.

If it is the case that pygmy chimpanzees are better at comprehending human language than at producing it (recall that in humans, language comprehension antedates its production; Benedict, 1979) and that several animal species process speech sounds categorically, as humans do (Kluender, Diehl, and Killeen, 1987), the consequences of these achievements need to be adequately evaluated. As for the latter point, Snowdon (1990) claims that the type of perception (categorical or not) can be induced by the set of stimuli used and by the instructions given to the subjects. Consequently, this perception "may be simply an irrelevant artifact that does not contribute to an understanding of the evolution of human speech and language" (Snowdon, 1990, p. 235). As for the former point, the impressive capacities shown by the pygmy chimpanzee might be difficult to interpret in evolutionary terms, unless we assume that comprehension has evolved in the absence of language. In that case, we may question the adaptive function of comprehension in the absence of production. This problem is reminiscent of the fact that tools are used by captive cebus monkeys but very rarely by wild cebus (see Chapter 4). These issues raise the problem of the role of the human-defined environment on the emergence of cognitive skills (for example, tool use and language) that have no equivalent in spontaneous behaviors and on which no selective pressure has supposedly acted.

In this respect, important gains could be made if ecologically oriented scientists studying problem solving in animals in their natural environments agree to cooperate with the generalists. It is cooperation that will strengthen the search for the significance of adaptive behaviors and the dimensions of the cognitive processes that control them.

Conclusions

Starting from the idea that animals and humans are constantly confronted with changing and complex arrays of external stimuli, Wasserman (1993) has argued that an efficient way to face the challenge of "reducing the demands of an organism's sensory and information-processing systems is for it to treat stimuli as members of a single class; by so doing, substantial cognitive economy can be achieved, thus freeing its adaptive machinery to deal with other competing exigencies of survival" (p. 216). An additional advantage to processing information categorically is to permit "an organism to identify novel stimuli as members of a particular class and to generalize knowledge about that category to these new members. Thus, an organism need not be bound to respond to only those stimuli with which it has had prior experience, correspondingly enhancing its ability to cope with a continually changing world" (p. 216). The evidence that categorical processing of stimuli is adaptive once again demonstrates the relevance of cognitive treatments of information with all the associate features (such as flexibility and generalization; see Chapter 1).

Another potential benefit of comparative animal cognition research depends on the willingness of its proponents to undertake systematic comparisons between very different species of varying phylogenetic distance from the human species. "By considering also species not closely related to humans, it is easier to overcome the danger of projecting the contents of our minds onto the experimental nonhuman animals. Furthermore, studying different species in their natural environment allows for the investigation of the adaptive value of cognition" (Prato Previde et al., 1992, pp. 97–98).

Moreover, the study of animal cognition—whether the generalist or ecological approach is adopted—should become an integral component of the science of cognition. Several benefits can be expected from giving full membership to animal cognition in the club of the cognitive sciences (which already includes neuroscience, cognitive psychology, philosophy, artificial intelligence, robotics, and linguistics). The main reason for including the study of animal cognition is that "to reach an integrated view of cognition, both developmental and evolutionary aspects are essential. Animal cognition contributes to the former, and is indeed crucial for the latter. Moreover, by studying cognitive processes in an ethological perspective, research on animals may shed light on the coupling between a cognitive system and its environment, thus introducing into cognitive science an ecological component that is still lacking" (Prato Previde et al., 1992, p. 98).

Finally, we should not forget the importance of interpretations drawn from observations of and experiments on living nonhuman species, particularly nonhuman primates. It is possible that the available data from comparative psychology may serve as a bedrock for a possible model of the human past, but in no way do they solve every mystery, because (1) today's nonhuman primates are not what our ancestors were and (2) these living primates are also the product of their own evolution. Keeping in mind this constraint, we may still argue that further study of animal cognition should continue to help us to understand our own intelligence and its evolution.

References

Index

References

Amsel, A., and M. E. Rashotte, eds. 1984. *Mechanisms of adaptive behavior: Clark L. Hull's theoretical papers, with commentary.* New York: Columbia University Press.

Anderson, J. R. 1984a. Monkeys with mirrors: Some questions for primate psychology. *International Journal of Primatology,* 5, 81–98.

―――― 1984b. The development of self-recognition: A review. *Developmental Psychobiology,* 17, 35–49.

―――― 1985. Development of tool-use to obtain food in a captive group of *Macaca tonkeana. Journal of Human Evolution,* 14, 637–645.

―――― 1986. Mirror-mediated finding of hidden food by monkeys *(Macaca tonkeana* and *Macaca fascicularis). Journal of Comparative Psychology,* 100, 237–242.

―――― 1990. Use of objects as hammers to open nuts by capuchin monkeys *(Cebus apella). Folia Primatologica,* 54, 138–145.

Anderson, J. R., and J.-J. Roeder. 1989. Responses of capuchin monkeys *(Cebus apella)* to different conditions of mirror-image stimulation. *Primates,* 30, 581–587.

Asendorpf, J. B., and P.-M. Baudonnière. 1993. Self-awareness and other-awareness: Mirror self-recognition and synchronic imitation among unfamiliar peers. *Developmental Psychology,* 29, 88–95.

Balda, R. P., and A. C. Kamil. 1989. A comparative study of cache recovery by three corvid species. *Animal Behaviour,* 38, 486–495.

Bard, K. A. 1990. "Social tool use" by free-ranging orangutans: A Piagetian and developmental perspective on the manipulation of an inanimate object. In S. T. Parker and K. R. Gibson, eds., *"Language" and intelligence in monkeys and apes: Comparative developmental perspectives,* pp. 356–378. New York: Cambridge University Press.

Bard, K. A., K. A. Platzman, B. M. Lester, and S. J. Suomi. 1992. Orientation to social and nonsocial stimuli in neonatal chimpanzees and humans. *Infant Behavior and Development,* 15, 43–56.

Bard, K. A., and J. Vauclair. 1984. The communicative context of object manipulation in ape and human adult-infant pairs. *Journal of Human Evolution,* 13, 181–190.

Baron-Cohen, S. 1992. How monkeys do things with "words." *The Behavioral and Brain Sciences,* 15, 148–149.

Bates, E. 1979. *The emergence of symbols: Cognition and communication in infancy.* New York: Academic Press.

Bates, E., L. Camaioni, and V. Volterra. 1975. The acquisition of performatives prior to speech. *Merril-Palmer Quarterly,* 21, 205–26.

Beck, B. B. 1972. Tool use in captive hamadryas baboons. *Primates,* 13, 276–296.

——— 1980. *Animal tool behavior: The use and manufacture of tools by animals.* New York: Garland.

Benedict, H. 1979. Early lexical development: Comprehension and production. *Journal of Child Language,* 6, 183–200.

Beninger, R. J., S. B. Kendall, and C. H. Vanderwolf. 1974. The ability of rats to discriminate their own behaviors. *Canadian Journal of Psychology,* 28, 79–91.

Bertrand, M. 1969. *The behavioral repertoire of the stumptail macaque.* Basel: Karger.

Beugnon, G., and R. Campan. 1989. Spatial memory and spatial cognition in insects. *Etologia,* 1, 63–86.

Bhatt, R. S., E. A. Wasserman, W. F. Reynolds, Jr., and K. S. Knauss. 1988. Conceptual behavior in pigeons: Categorization of both familiar and novel examples from four classes of natural and artificial stimuli. *Journal of Experimental Psychology: Animal Behavior Processes,* 14, 219–234.

Bickerton, D. 1990. *Language and species.* Chicago: Chicago University Press.

Boakes, R. 1984. *From Darwin to behaviorism.* New York: Cambridge University Press.

Boesch, C. 1991. Teaching among wild chimpanzees. *Animal Behaviour,* 41, 530–532.

——— 1993a. Aspects of transmission of tool-use in wild chimpanzees. In K. R. Gibson and T. Ingold, eds., *Tools, language and cognition in human evolution,* pp. 171–193. New York: Cambridge University Press.

——— 1993b. Towards a new image of culture in wild chimpanzees. *The Behavioral and Brain Sciences,* 16, 514–515.

Boesch, C., and Boesch, H. 1983. Optimization of nut-cracking with natural hammers by wild chimpanzees. *Behaviour,* 83, 265–286.

——— 1984. Mental maps in wild chimpanzees: An analysis of hammer transports for nut cracking. *Primates,* 25(2), 160–170.

——— 1990. Tool use and tool making in wild chimpanzees. *Folia Primatologica,* 54, 86–99.

Boesch, C., P. Marchesi, N. Marchesi, B. Fruth, and F. Joulian. 1994. Is nut cracking in wild chimpanzees a cultural behavior? *Journal of Human Evolution*, 26, 325–338.

Boinski, S. 1988. Use of a club by a white-faced capuchin *(Cebus capucinus)* to attack a venomous snake *(Bothrops asper)*. *American Journal of Primatology*, 14, 177–179.

Bolhuis, H. J., and H. S. van Kampen. 1988. Serial position curves in spatial memory of rats: Primacy and recency effects. *Quarterly Journal of Experimental Psychology*, 40B, 135–159.

Boysen, S. T., and G. B. Berntson. 1989. Numerical competence in a chimpanzee *(Pan troglodytes)*. *Journal of Comparative Psychology*, 103, 23–31.

Boysen, S. T., G. B. Berntson, and K. S. Quigley. 1993. Processing of ordinality by chimpanzees *(Pan troglodytes)*. *Journal of Comparative Psychology*, 107, 208–215.

Boysen, S. T., G. B. Berntson, T. A. Shreyer, and M. B. Hannan. 1995. Indicating acts during counting by a chimpanzee *(Pan troglodytes)*. *Journal of Comparative Psychology*, 109, 47–51.

Boysen, S. T., and E. J. Capaldi, eds. 1993. *The development of numerical competence: Animal and human models*. Hillsdale, NJ: Lawrence Erlbaum Associates.

Brainerd, C. J. 1978. The stage question in cognitive developmental theory. *The Behavioral and Brain Sciences*, 1, 173–213.

Bronstein, P. M. 1983. Agonistic sequences and the assessment of opponents in male *Betta splendens*. *American Journal of Psychology*, 96, 163–177.

Brooks-Gunn, J., and M. Lewis. 1984. The development of early visual self-recognition. *Developmental Review*, 4, 215–239.

Bruner, J. S. 1983. *Child talk*. New York: Norton.

Burghardt, G. M. 1977. Learning processes in reptiles. *Biology of the Reptilia*, 7, 555–681.

Byrne, R. W., and A. Whiten. 1988. *Machiavellian intelligence*. Oxford: Clarendon Press.

Candland, D. K., J. A. French, and C. N. Johnson. 1978. Object play: A test for categorized model by the genesis of object-play in *Macaca fuscata*. In E. O. Smith, ed., *Social play in primates*, pp. 259–296. New York: Academic Press.

Carpenter, C. 1934. A field study of the behavior and social relations of howling monkeys. *Comparative Psychology Monographs*, 10(2), 1–168.

Cartwright, B. A., and T. S. Collett. 1983. Landmark learning in bees: Experiments and models. *Journal of Comparative Physiology*, 151, 521–543.

——— 1987. Landmark maps for honeybees. *Biological Cybernetics*, 57, 85–93.

Castro, C. A., and T. Larsen. 1992. Primacy and recency effects in nonhuman primates. *Journal of Comparative Psychology*, 18, 335–340.

Chapman, C. A. 1986. Boa constrictor predation and group response in white-faced cebus monkeys. *Biotropica*, 18, 171–172.

Cheney, D. L., and R. M. Seyfarth. 1980. Vocal recognition in free-ranging vervet monkeys. *Animal Behaviour,* 28, 362–367.

——— 1982. How vervet monkeys perceive their grunts: Field playback experiments. *Animal Behaviour,* 30, 739–751.

——— 1985. The social and non-social world of non-human primates. In R. A. Hinde, A.-N. Perret-Clermont, and J. Stevenson-Hinde, eds., *Social relationships and cognitive development,* pp. 23–44. Oxford: Clarendon Press.

——— 1990a. *How monkeys see the world: Inside the mind of another species.* Chicago: University of Chicago Press.

——— 1990b. The representation of social relations by monkeys. *Cognition,* 37, 167–196.

——— 1990c. Attending to behaviour versus attending to knowledge: Examining monkeys' attribution of mental states. *Animal Behaviour,* 40, 742–753.

——— 1992. Precis of "How monkeys see the world." *The Behavioral and Brain Sciences,* 15, 135–182.

Cheney, D. L., R. M. Seyfarth, and B. Smuts. 1986. Social relationships and social cognition in nonhuman primates. *Science,* 234, 1361–1366.

Cheng, K. 1986. A purely geometric module in the rat's spatial memory. *Cognition,* 23, 149–178.

Chevalier-Skolnikoff, S., and J. Liska. 1993. Tool use by wild and captive elephants. *Animal Behaviour,* 46, 209–219.

Chomsky, N. 1968. *Language and mind.* New York: Harcourt, Brace and World.

Church, R. M. 1993. Human models of animal cognition. *Psychological Science,* 4, 170–173.

Church, R. M., and J. Gibbon. 1982. Temporal generalization. *Journal of Experimental Psychology: Animal Behavior Processes,* 8, 165–186.

Corballis, M. C. 1988. Recognition of disoriented shapes. *Psychological Review,* 95, 115–123.

Corballis, M. C., and I. L. Beale. 1976. *The psychology of left and right.* Hillsdale, NJ: Lawrence Erlbaum Associates.

Custance, D., and K. A. Bard. 1994. The development of gestural imitation and self-recognition in chimpanzees *(Pan troglodytes).* In S. T. Parker, R. W. Mitchell, and M. L. Boccia, eds., *Self-awareness in animals and humans: Developmental perspectives,* pp. 207–226. New York: Cambridge University Press.

Couvillon, P. A., and M. E. Bitterman. 1992. A conventional conditioning analysis of "transitive inference" in pigeons. *Journal of Experimental Psychology: Animal Behavior Processes,* 18, 308–310.

Crystal, J. D., and S. J. Shettleworth. 1994. Spatial list learning in black-capped chickadees. *Animal Learning and Behavior,* 22, 77–83.

D'Amato, M. R., and M. Colombo. 1988. Representation of serial order in monkeys *(Cebus apella). Journal of Experimental Psychology: Animal Behavior Processes,* 14, 131–139.

D'Amato, M. R., D. P. Salmon, and M. Colombo. 1985. Extent and limits of the matching concepts in monkeys *(Cebus apella)*. *Journal of Experimental Psychology: Animal Behavior Processes,* 11, 35–51.

Darwin, C. 1871. *The descent of man and selection in relation to sex.* London: Murray.

Dasser, V. 1985. Complexity in primate social relationships. In R. A. Hinde, A. Perret-Clermont, and J. Stevenson-Hinde, eds., *Social relationships and cognitive development,* pp. 3–22. Oxford: University Press.

———— 1987a. Slides of group members as representations of real animals *(Macaca fascicularis)*. *Ethology,* 76, 65–73.

———— 1987b. *A social concept of monkeys.* Inaugural dissertation, University of Zürich.

———— 1988. A social concept in Java monkeys. *Animal Behaviour,* 36, 225–230.

Davis, H. 1992. Transitive inference in rats *(Rattus norvegicus)*. *Journal of Comparative Psychology,* 106, 342–349.

Davis, H., and S. A. Bradford. 1986. Counting behavior by rats in a simulated natural environment. *Ethology,* 73, 265–280.

Davis, H., and R. Pérusse. 1988. Numerical competence in animals: Definitional issues, current evidence, and a new research agenda. *The Behavioral and Brain Sciences,* 11, 561–615.

Dawson, B. V., and B. M. Foss. 1965. Observational learning in budgerigars. *Animal Behaviour,* 13, 470–474.

Deag, J. 1977. Aggression and submission in monkey societies. *Animal Behaviour,* 25, 465–474.

Deag, J., and J. Crook. 1971. Social behaviour and "agonistic buffering" in the wild Barbary macaque. *Folia Primatologica,* 15, 183–200.

De Blois, S. T., and M. A. Novak. 1994. Object permanence in rhesus monkeys *(Macaca mulatta)*. *Journal of Comparative Psychology,* 108, 318–327.

De Saussure, F. 1966. *Course in general linguistics.* New York: McGraw-Hill. [First published in French in 1916.]

De Waal, F. B. M. 1982. *Chimpanzee politics.* London: Jonathan Cape.

———— 1989. *Peacemaking among primates.* Cambridge, MA: Harvard University Press.

———— 1991. Complementary methods and convergent evidence in the study of primate cognition. *Behaviour,* 118, 297–320.

Dickinson, A. 1980. *Contemporary animal learning theory.* New York: Cambridge University Press.

Doré, F. Y. 1986. Object permanence in adult cats *(Felix catus)*. *Journal of Comparative Psychology,* 100, 340–347.

Doré, F. Y., and C. Dumas. 1987. Psychology of animal cognition: Piagetian studies. *Psychological Bulletin,* 102, 219–233.

Dumas, C. 1992. Object permanence in cats *(Felix catus)*: An ecological approach to the study of invisible displacements. *Journal of Comparative Psychology,* 106, 404–410.

Dyer, F. C. 1991. Bees acquire route-based memories but not cognitive maps in a familiar landscape. *Animal Behaviour,* ˙41, 239–246.

Eibl-Eibesfeldt, I. 1970. *Ethology: The biology of behavior.* New York: Holt, Rinehart and Winston.

Epstein, R. 1982. "Representation" in the chimpanzee. *Psychological Report,* 50, 745–746.

Epstein, R., R. P. Lanza, and B. F. Skinner. 1981. "Self-awareness" in the pigeon. *Science,* 212, 695–696.

Etienne, A. S. 1973a. Developmental stages and cognitive structures as determinants of what is learned. In R. A. Hinde and J. Stevenson-Hinde, eds., *Constraints on learning,* pp. 371–395. London: Academic Press.

——— 1973b. Searching behaviour towards a disappearing prey in the domestic chick as affected by preliminary experience. *Animal Behaviour,* 21, 149–161.

——— 1984. The meaning of object permanence at different zoological levels. *Human Development,* 27, 309–320.

Evans, C. S., L. Evans, and P. Marler. 1993. On the meaning of alarm calls: Functional reference in an avian vocal system. *Animal Behaviour,* 46, 23–38.

Fernandes, M. E. B. 1991. Tool use and predation of oysters *(Crassostrea rhizophorea)* by the tufted capuchin, *Cebus apella apella,* in Brackish Water Mangrove Swamp. *Primates,* 32, 529–531.

Field, T. 1979. Differential behavioral and cardiac responses of 3-month-old infants to a mirror and peer. *Infant Behavior and Development,* 2, 179–184.

Fiorito, G., and P. Scotto. 1992. Observational learning in *Octopus vulgaris. Science,* 256, 545–547.

Flavell, J. H. 1963. *The developmental psychology of Jean Piaget.* New York: Van Nostrand.

Fobes, J. L., and J. E. King. 1982. Measuring primate learning abilities. In J. L. Fobes and J. E. King, eds., *Primate behavior,* pp. 289–326. New York: Academic Press.

Fodor, J. A. 1975. *The language of thought.* Cambridge, MA: Harvard University Press.

——— 1983. *The modularity of mind.* Cambridge, MA: MIT Press.

Frank, H., and M. G. Frank. 1985. Comparative manipulation-test performance in ten-week-old wolves *(Canis lupus)* and Alaskan malamutes *(Canis familiaris):* A Piagetian interpretation. *Journal of Comparative Psychology,* 99, 266–274.

Gaffan, E. A. 1992. Primacy, recency, and the variability of data in studies of animals' working memory. *Animal Learning and Behavior,* 20, 240–252.

——— 1994. Primacy in animals' working memory: Artifacts. *Animal Learning and Behavior,* 22, 231–232.

Gagliardi, J. L., K. K. Kirkpatrick-Steger, J. Thomas, G. J. Allen, and M. S. Blumberg.

1995. Seeing and knowing: Knowledge attribution versus stimulus control in adult humans *(Homo sapiens). Journal of Comparative Psychology,* 109, 107–124.

Gagnon, S., and F. Y. Doré. 1994. Cross-sectional study of object permanence in domestic puppies *(Canis familiaris). Journal of Comparative Psychology,* 108, 220–232.

Galef, B. G. 1987. Comparative psychology is dead! Long life to comparative psychology. *Journal of Comparative Psychology,* 101, 259–261.

——— 1988. Imitation in animals: History, definitions, and interpretation of data from the psychological laboratory. In T. Zentall and B. G. Galef, eds., *Social learning: Psychological and biological perspectives,* pp. 3–28. Hillsdale, NJ: Lawrence Erlbaum Associates.

——— 1990. Tradition in animals: Field observations and laboratory analyses. In M. Bekoff and D. Jamieson, eds., *Interpretation and explanation in the study of animal behavior,* vol. 1, pp. 74–95. Boulder, CO: Westview Press.

——— 1992. The question of animal culture. *Human Nature,* 3, 157–178.

Galef, B. G., L. A. Manzig, and R. M. Field. 1986. Imitation learning in budgerigars: Dawson and Foss 1965 revisited. *Behavioural Processes,* 13, 191–202.

Gallistel, C. R. 1989. Animal cognition: The representation of space, time and number. *Annual Review of Psychology,* 40, 155–189.

——— 1990. *The organization of learning.* Cambridge, MA: MIT Press.

Gallup, G. G. 1970. Chimpanzees: Self-recognition. *Science,* 167, 86–87.

——— 1975. Toward an operational definition of self-awareness. In R. H. Tuttle, ed., *Socioecology and psychology of primates,* pp. 309–341. The Hague, Netherlands: Mouton.

——— 1983. Toward a comparative psychology of mind. In R. L. Mellgren, ed., *Animal cognition and behavior,* pp. 473–510. Amsterdam: North-Holland.

Gardner, H. 1985. *The mind's new science: A history of the cognitive revolution.* New York: Basic Books.

Gardner, R. A., and B. T. Gardner. 1969. Teaching sign language to a chimpanzee. *Science,* 165, 664–672.

——— 1984. A vocabulary test for chimpanzees *(Pan troglodytes). Journal of Comparative Psychology,* 98, 381–404.

Gardner, R. A., B. T. Gardner, and T. E. Van Cantfort. 1989. *Teaching sign language to chimpanzees.* Albany: State University of New York Press.

Gelman, R., and C. R. Gallistel. 1978. *The child's understanding of number.* Cambridge, MA: Harvard University Press.

Gibbon, J., and R. M. Church. 1984. Sources of variance in an information processing theory of timing. In H. L. Roitblat, T. G. Bever, and H. S. Terrace, eds., *Animal cognition,* pp. 465–488. Hillsdale, NJ: Lawrence Erlbaum Associates.

Gibson, K. R. 1986. Cognition, brain size and the extraction of embedded food resources. In J. Else and P. Lee, eds., *Primate ontogeny, cognition and social behaviour*, pp. 93–103. New York: Cambridge University Press.

Gibson, K. R., and T. Ingold, eds. 1993. *Tools, language and cognition in human evolution.* New York: Cambridge University Press.

Gillan, D. D. 1981. Reasoning in the chimpanzee. II. Transitive inference. *Journal of Experimental Psychology: Animal Behavior Processes*, 7, 150–164.

Gillan, D. D., D. Premack, and G. Woodruff. 1981. Reasoning in the chimpanzee: I. Analogical reasoning. *Journal of Experimental Psychology: Animal Behavior Processes*, 7, 1–17.

Gisiner, R., and R. J. Schusterman. 1992. Sequence, syntax, and semantics: Responses of a language trained sea lion *(Zalophus californianus)* to novel sign combinations. *Journal of Comparative Psychology*, 106, 78–91.

Glickman, S. E., and R. W. Sroges. 1966. Curiosity in zoo animals. *Behaviour*, 26, 151–188.

Gomez, J. C. 1991. Visual behaviour as a window for reading the mind of others in primates. In A. Whiten, ed., *Natural theories of mind*, pp. 195–207. Oxford: Basil Blackwell.

Goodall, J. 1968. The behaviour of free-living chimpanzees in the Gombe Stream Reserve Tanzania. *Animal Behaviour Monographs*, 1, 161–311.

——— 1971. *In the shadow of man.* Boston: Houghton Mifflin.

——— 1986. *The chimpanzees of Gombe.* Cambridge, MA.: Harvard University Press.

Gordon, C. W. 1983. The malleability of memory in animals. In R. L. Mellgren, ed., *Animal cognition and behavior*, pp. 399–426. Amsterdam: North-Holland.

Gould, J. L. 1986. The locale map of honey-bee: Do insects have cognitive maps? *Science*, 232, 861–863.

Gould, S. J. 1977. *Ontogeny and phylogeny.* Cambridge, MA: Harvard University Press.

Gouzoules, S., H. Gouzoules, and P. Marler. 1984. Rhesus monkey *(Macaca mulatta)* screams: Representational signalling in the recruitment of agonistic aid. *Animal Behaviour*, 32, 182–193.

Gray, P. H. 1973. Comparative psychology and ethology: A saga of twins reared apart. In E. Tobach, H. E. Adler, and L. L. Adler, eds., *Comparative psychology at issue. Annals of the New York Academy of Sciences*, 223, 49–53.

Greenfield, P. M. 1991. Language, tool, and brain: The ontogeny and phylogeny of hierarchically organized sequential behavior. *The Behavioral and Brain Sciences*, 14, 521–595.

Greenfield, P. M., and E. S. Savage-Rumbaugh. 1990. Grammatical combination in *Pan paniscus*: Processes of learning and invention in the evolution and development of language. In S. T. Parker and K. R. Gibson, eds., *"Language" and intelligence in monkeys and apes: Comparative developmental perspectives*, pp. 540–578. New York: Cambridge University Press.

Griffin, D. R. 1976. *The question of animal awareness.* New York: Rockefeller University Press.

———— 1978. Prospects for a cognitive ethology. *The Behavioral and Brain Sciences,* 1, 527–538.

———— ed. 1982. *Animal mind—human mind.* New York: Springer Verlag.

———— 1984. *Animal thinking.* Cambridge, MA: Harvard University Press.

———— 1991. Progress toward a cognitive ethology. In C. A. Ristau, ed., *Cognitive ethology: The minds of other animals,* pp. 3–17. Hillsdale: Lawrence Erlbaum Associates.

———— 1992. *Animal minds.* Chicago: University of Chicago Press.

Gruber, H. E., Girgus, J. S., and Banuazizi, A. 1971. The development of object permanence in the cat. *Developmental Psychology,* 14, 9–15.

Guillaume, P. 1940. *La Psychologie animale.* Paris: Colin.

Hall, K., and G. B. Schaller. 1964. Tool using behavior of the California sea otter. *Journal of Mammalogy,* 45, 287–298.

Hamilton, C. R., S. B. Tieman, and B. A. Brody. 1973. Interhemispheric comparison of mirror-image stimuli in chiasm-sectioned monkeys. *Brain Research,* 58, 415–425.

Hamilton, W., R. Buskirk, and W. Buskirk. 1975. Defensive stoning by baboons. *Nature,* 256, 488–489.

Harcourt, A. H., and F. B. M. de Waal, eds. 1992. *Coalitions and alliances in humans and other animals.* Oxford: Oxford University Press.

Harlow, H. F. 1949. The formation of learning sets. *Psychological Review,* 56, 51–65.

Hayes, C. 1951. *The ape in our house.* New York: Harper.

Hayes, K. J., and C. H. Hayes. 1952. Imitation in a home-raised chimpanzee. *Journal of Comparative and Physiological Psychology,* 45, 450–459.

Hemelrijk, C. K. 1994. Support for being groomed in long-tailed macaques, *Macaca fascicularis. Animal Behaviour,* 48, 479–481.

Herman, L. M. 1986. Cognition and language competencies of bottlenosed dolphins. In R. J. Schusterman, J. A. Thomas, and F. G. Wood, eds., *Dolphin behavior and cognition: Comparative and ecological aspects.* pp. 221–252. Hillsdale, NJ: Lawrence Erlbaum Associates.

Herman, L. M., and J. A. Gordon. 1974. Auditory delayed matching in the bottlenosed dolphin. *Journal of the Experimental Analysis of Behavior,* 21, 19–26.

Herman, L. M., D. G. Richards, and J. P. Wolz. 1984. Comprehension of sentences by bottlenosed dolphins. *Cognition,* 16, 129–219.

Herman, L. M., A. A. Pack, and P. Morrel-Samuels. 1993. Representational and conceptual skills of dolphins. In H. L. Roitblat, L. M. Herman, and P. E. Nachtigall, eds., *Language and communication: Comparative perspectives,* pp. 403–442. Hillsdale, NJ: Lawrence Erlbaum Associates.

Herrnstein, R. J. 1979. Acquisition, generalization, and discrimination reversal of a natural concept. *Journal of Experimental Psychology: Animal Behavior Processes,* 7, 150–164.

―――― 1990. Levels of stimulus control: A functional approach. *Cognition*, 37, 133–166.

Herrnstein, R. J., D. H. Loveland, and C. Cable. 1976. Natural concepts in pigeons. *Journal of Experimental Psychology: Animal Behavior Processes*, 2, 285–311.

Herrnstein, R. J., and P. A. de Villiers. 1980. Fish as a natural category for people and pigeons. In G. H. Bower, ed., *The psychology of learning and motivation*, vol. 14, pp. 60–97. New York: Academic Press.

Hewes, G. W. 1973. Primate communication and the gestural origin of language. *Current Anthropology*, 14, 5–24.

Heyes, C. M. 1993a. Imitation, culture and cognition. *Animal Behaviour*, 46, 999–1010.

―――― 1993b. Anecdotes, trapping and triangulating: Do animals attribute mental states? *Animal Behaviour*, 46, 177–188.

―――― 1993c. Reflections on self-recognition in primates. *Animal Behaviour*, 47, 909–919.

―――― 1994. Social learning in animals: Categories and mechanisms. *Biological Review*, 69, 207–231.

Heyes, C. M., and G. R. Dawson. 1990. A demonstration of observational learning using a bidirectional control. *Quarterly Journal of Experimental Psychology*, 42B, 59–71.

Heyes, C. M., and G. R. Dawson, and T. Nokes. 1992. Imitation in rats: Initial responding and transfer evidence. *Quarterly Journal of Experimental Psychology*, 45B, 229–240.

Heyes, C. M., E. Jaldow, T. Nokes, and G. R. Dawson. 1994. Imitation in rats *(Rattus norvegicus):* The role of demonstrator action. *Behavioural Processes*, 32, 173–182.

Hinde, R. A. 1982. *Ethology*. New York: Oxford University Press.

Hinde, R. A., and J. Stevenson-Hinde, eds. 1973. *Constraints on learning: Limitations and predispositions*. New York: Academic Press.

Hockett, C. F. 1960. The origin of speech. *Scientific American*, 203, 88–96.

Holland, P. C., and J. J. Straub. 1979. Differential effects of two ways of devaluating the unconditioned stimulus after Pavlovian appetitive conditioning. *Journal of Experimental Psychology: Animal Behavior Processes*, 5, 178–193.

Hollard, V. D., and J. D. Delius. 1982. Rotational invariance in visual pattern recognition by pigeons and humans. *Science*, 218, 804–806.

Hopkins, W. D., J. Fagot, and J. Vauclair. 1993. Mirror-image matching and mental rotation problem solving by baboons *(Papio papio):* Unilateral input enhances performance. *Journal of Experimental Psychology: General*, 122, 61–72.

Hull, C. L. 1934. The concept of the habit-family hierarchy and maze learning. *Psychological Review*, 41, 33–54.

Hulse, S. H., H. Fowler, and W. K. Honig, eds. 1978. *Cognitive processes in animal behavior*. Hillsdale, NJ: Lawrence Erlbaum Associates.

Humphrey, N. 1976. The social function of intellect. In P. P. G. Bateson and R. A.

Hinde, eds., *Growing points in ethology*, pp. 303–317. New York: Cambridge University Press.

Hunter, W. S. 1912. The delayed reaction in animals. *Behavioral Monographs*, 2, 1–85.

Ianco-Worrall, A. D. 1972. Bilingualism and cognitive development. *Child Development*, 43, 1390–1400.

Itakura, S. 1987. Mirror guided behavior in Japanese monkeys *(Macaca fuscata fuscata)*. *Primates*, 28, 149–161.

Izawa, K. 1979. Foods and feeding behavior of wild black-capped capuchins *(Cebus apella)*. *Primates*, 20, 57–76.

Jennings, H. S. 1906. *The behavior of the lower organisms.* New York: Macmillan.

Jerison, H. J. 1973. *Evolution of the brain and intelligence.* New York: Academic Press.

Johnson-Laird, P. N. 1988. *The computer and the mind: An introduction to cognitive science.* Cambridge, MA: Harvard University Press.

Jolly, A. 1966. Lemur social behavior and primate intelligence. *Science*, 153, 501–506.

Jorgensen, M. J. 1994. *Investigations of the antecedents of self-recognition using a video-task paradigm in capuchins (Cebus apella) and chimpanzees (Pan troglodytes).* Dissertation, University of California, Riverside.

Joubert, A., and J. Vauclair. 1986. Reaction to novel objects in a troop of Guinea baboons: Approach and manipulation. *Behaviour*, 96, 92–104.

Kamil, A. C. 1978a. A synthetic approach to the study of animal intelligence. In D. W. Leger, ed., *Nebraska Symposium on Motivation*, vol. 35: *Comparative perspectives in modern psychology*, pp. 257–308. Lincoln: University of Nebraska Press.

——— 1978b. Systematic foraging by a nectar-feeding bird, the Amakiki *(Loxops virens)*. *Journal of Comparative and Physiological Psychology*, 92, 388–396.

——— 1984. Adaptation and cognition: Knowing what comes naturally. In H. L. Roitblat, T. G. Bever, and H. S. Terrace, eds., *Animal cognition*, pp. 533–544. Hillsdale: Lawrence Erlbaum Associates.

Kamil, A. C., and R. P. Balda. 1985. Cache recovery and spatial memory in Clark's nutcrackers *(Nucifraga columbiana)*. *Journal of Experimental Psychology: Animal Behavior Processes*, 11, 95–111.

Kawai, M. 1965. Newly acquired precultural behavior of the natural troop of Japanese monkeys on Koshima Islet. *Primates*, 6, 1–30.

Keeley, E. R., and J. W. Grant. 1993. Visual information, resource value and sequential assessment in convict cichlid *(Cichlasoma nigrofasciatum)* contests. *Behavioral Ecology*, 4, 345–349.

Kesner, R. P., A. A. Chiba, and P. Jackson-Smith. 1994. Rats do show primacy and recency effects in memory for lists of spatial location: A reply to Gaffan. *Animal Learning and Behavior*, 22, 214–218.

Killeen, P. R., and G. Fetterman. 1988. A behavioral theory of timing. *Psychological Review,* 95, 274–295.

Kimura, D. 1979. Neuromotor mechanisms in the evolution of human communication. In H. D. Steklis and M. J. Raleigh, eds., *Neurobiology of social communication in primates: An evolutionary perspective,* pp. 197–219. New York: Academic Press.

Kirchner, W. H., and U. Braun. 1994. Dancing honey bees indicate the location of food sources using path integration rather than cognitive maps. *Animal Behaviour,* 48, 1437–1441.

Kluender, K. R., R. L. Diehl, and P. R. Killeen. 1987. Japanese quail can learn phonetic categories. *Science,* 237, 1195–1197.

Klüver, H. 1933. *Behavior mechanisms in monkeys.* Chicago: Chicago University Press.

Koehler, O. 1950. The abilities of birds to "count." *Bulletin of Animal Behaviour,* 9, 41–45.

Köhler, W. 1925. *The mentality of apes.* London: Routledge and Kegan Paul.

Konishi, M. 1963. The role of auditory feedback in the vocal behaviour of the domestic fowl. *Zeitschrift für Tierpsychologie,* 20, 349–367.

Kosslyn, S. M. 1980. *Image and mind.* Cambridge, MA: Harvard University Press.

Kummer, H. 1967. Tripartite relations in hamadryas baboons. In S. Altmann, ed., *Social communication among primates,* pp. 63–71. Chicago: Chicago University Press.

——— 1971. *Primate societies.* Chicago: Aldine Atherton.

——— 1982. Social knowledge in free-ranging primates. In D. R. Griffin ed., *Animal mind—human mind,* pp. 113–130. Berlin: Springer Verlag.

Kummer, H., W. Götz, and W. Angst. 1974. Triadic differentiation: An inhibitory process protecting pair bonds in baboons. *Behaviour,* 48, 62–87.

Kummer, H., V. Dasser, and P. Hoyningen-Huene. 1990. Exploring primate social cognition: Some critical remarks. *Behaviour,* 112, 84–98.

Lack, D. 1953. Darwin's finches. *Scientific American,* 188, 66–72.

Ledbetter, D. H., and J. A. Basen. 1982. Failure to demonstrate self-recognition in gorillas. *American Journal of Primatology,* 2, 307–310.

Leger, D. W., D. H. Owings, and L. M. Boal. 1979. Contextual information and differential responses to alarm whistles in California ground squirrels. *Zeitschrift für Tierpsychologie,* 49, 142–155.

Leslie, A. M. 1987. Pretense and representation: Origins of "theory of mind." *Psychological Review,* 94, 412–426.

Leslie, A. M., and U. Firth. 1988. Autistic children's understanding of seeing, knowing, and believing. *British Journal of Developmental Psychology,* 6, 315–324.

Lethmate, J., and G. Dücker. 1973. Untersuchungen zum Selbsterkennen im Spiegel bei Orang-utans und einigen anderen Affenarten. *Zeitschrift für Tierpsychologie,* 33, 248–269.

Lewis, M., and J. Brooks-Gunn. 1979. *Social cognition and the acquisition of self.* New York: Plenum Press.

Lieberman, P. 1975. *On the origins of language.* New York: Macmillan.

—— 1984. *The biology and evolution of language.* Cambridge, MA: Harvard University Press.

Loeb, J. 1900. *Comparative physiology of the brain and comparative psychology.* New York: Putnam.

MacDonald, S. E. 1994. Gorillas' *(Gorilla gorilla gorilla)* spatial memory in a foraging task. *Journal of Comparative Psychology,* 108, 107–113.

MacDonald, S. E., and D. M. Wilkie. 1990. Yellow-nosed monkeys's *(Cercopithecus ascanius whitesidei)* spatial memory in a simulated foraging environment. *Journal of Comparative Psychology,* 104, 382–387.

MacDonald, S. E., J. C. Pang, and S. Gibeault. 1994. Marmoset *(Callithrix jacchus jacchus)* spatial memory in a foraging task: Win-stay versus win-shift strategies. *Journal of Comparative Psychology,* 108, 328–334.

Macphail, E. M. 1982. *Brain and intelligence in vertebrates.* Oxford: Clarendon Press.

—— 1987. The comparative psychology of intelligence. *The Behavioral and Brain Sciences,* 10, 645–696.

—— 1990. Continuing commentary to "The comparative psychology of intelligence." *The Behavioral and Brain Sciences,* 13, 391–398.

—— 1993. *The neuroscience of animal intelligence: From the seahare to the seahorse.* New York: Columbia University Press.

Maier, N. R. F., and T. C. Schneirla. 1935. *Principles of animal psychology.* New York: MacGraw-Hill.

Manning, A., and M. S. Dawkins. 1992. *An Introduction to animal behaviour.* New York: Cambridge University Press.

Margules, J., and C. R. Gallistel. 1988. Heading in the rat: Determination by an environmental shape. *Animal Learning and Behavior,* 16, 404–410.

Marler, P., A. Dufty, and R. Pickert. 1986a. Vocal communication in the domestic chicken: I. Does a sender communicate information about the quality of a food referent to a receiver? *Animal Behaviour,* 34 188–193.

—— 1986b. Vocal communication in the domestic chicken: II. Is a sender sensitive to the presence and nature of a receiver? *Animal Behaviour,* 34, 194–198.

Mason, W. A. 1976. Windows on other minds. *Science,* 194, 930–931.

Mathieu, M., M. A. Bouchard, L. Granger, and J. Herscovitch. 1976. Piagetian object permanence in *Cebus capucinus, Lagothrica flavicauda* and *Pan troglodytes. Animal Behaviour,* 24, 585–588.

Mathieu, M., N. Daudelin, Y. Dagenais, and T. Gouin Décarie. 1980. Piagetian causality in two house-reared chimpanzees *(Pan troglodytes). Canadian Journal of Psychology,* 34, 179–186.

Mathieu, M., and G. Bergeron. 1981. Piagetian assessment on cognitive development in chimpanzees *(Pan troglodytes)*. In A. B. Chiarelli and R. S. Coruccini, eds., *Primate behavior and sociobiology*, pp. 142–147. New York: Springer-Verlag.

Matsuzawa, T. 1985. Use of numbers by a chimpanzee. *Nature*, 315, 57–59.

———— 1987. Color naming and classification in a chimpanzee *(Pan troglodytes)*. *Journal of Human Evolution*, 14, 283–291.

———— 1991. Use of numbers by a chimpanzee: A further study. In A. Ehara, T. Kimura, O. Takenara, and M. Iwamoto, eds., *Primatology today*, pp. 317–320. Amsterdam: Elsevier.

———— 1994. Field experiments on use of stone tools by chimpanzees in the wild. In R. W. Wrangham, W. C. McGrew, F. B. M. de Waal, and P. G. Heltne, eds., *Chimpanzee cultures*, pp. 351–370. Cambridge, MA: Harvard University Press.

McCrary, J. W., and W. S. Hunter. 1953. Serial position curves in verbal learning. *Science*, 117, 131–134.

McFarland, D. 1989. *Problems of animal behaviour*. New York: John Wiley and Sons.

McGonigle, B. O., and M. Chalmers. 1977. Are monkeys logical? *Nature*, 267, 694–696.

McGrew, W. C. 1977. Socialization and object manipulation of wild chimpanzees. In S. Chevalier-Skolnikoff and F. E. Poirier, eds., *Primate bio-social development*, pp. 261–288. New York: Garland.

———— 1992. *Chimpanzee material culture: Implications for human evolution*. New York: Cambridge University Press.

———— 1993. The intelligent use of tools: Twenty propositions. In K. R. Gibson and T. Ingold, eds., *Tools, language and cognition in human evolution*, pp. 151–170. New York: Cambridge University Press.

Meltzoff, A. N., and K. Moore. 1977. Imitation of facial and manual gestures by human neonates. *Science*, 198, 75–78.

Menzel, E. W. 1973. Chimpanzee spatial memory. *Science*, 182, 943–945.

———— 1978. Cognitive mapping in chimpanzees. In S. H. Hulse, H. Fowler, and W. K. Honig, eds., *Cognitive processes in animal behavior*, pp. 375–422. Hillsdale, NJ: Lawrence Erlbaum Associates.

———— 1987. Behavior as a locationist views it. In P. Ellen and C. Thinus-Blanc, eds., *Cognitive processes and spatial orientation in animal and man*, vol. 1: *Experimental psychology and ethology*, pp. 55–72. Dordrecht: Martinus Nijhoff.

Menzel, E. W., D. Premack, and G. Woodruff. 1978. Map reading by chimpanzees. *Folia Primatologica*, 29, 241–249.

Menzel, E. W., and C. R. Menzel. 1979. Cognitive, developmental and social aspects of responsiveness to novel objects in a family group of marmosets (*Saguinus fuscicolis*). *Behaviour*, 70, 251–279.

Menzel, E. W., E. S. Savage-Rumbaugh, and J. Lawson. 1985. Chimpanzee *(Pan troglodytes)* spatial problem solving with the use of mirrors and televised equivalents of mirrors. *Journal of Comparative Psychology*, 99, 211–217.

Merz, E. 1978. Male-male interactions with dead infants in *Macaca sylvanus*. *Primates*, 19, 749–754.

Mignault, C. 1985. Transition between sensori-motor and symbolic activities in nursery-reared chimpanzees *(Pan troglodytes)*. *Journal of Human Evolution*, 14, 747–758.

Miles, H. L. 1983. Apes and language: The search for communicative competence. In J. de Luce and H. T. Wilder, eds., *Language in apes*, pp. 43–61. New York: Springer Verlag.

———— 1990. The cognitive foundations for reference in a signing orangutan. In S. T. Parker and K. R. Gibson, eds., *"Language" and intelligence in monkeys and apes: Comparative developmental perspectives*, pp. 511–539. New York: Cambridge University Press.

Milton, K. 1981. Distribution patterns of tropical plant foods as an evolutionary stimulus to primate mental development. *American Anthropologist*, 83, 534–548.

Mitchell, R. W., and J. R. Anderson. 1993. Discrimination learning of scratching, but failure to obtain imitation and self-recognition in a long-tailed macaque. *Primates*, 34, 301–309.

Mitchell, R. W., and N. S. Thompson, eds. 1986. *Deception: Perspectives on human and nonhuman deceit*. Albany, NY: SUNY Press.

Moore, B. R. 1993. Avian movement imitation and a new form of mimicry: Tracing the evolution of a complex form of learning. *Behaviour Behaviour*, 122, 231–263.

Morgan, C. L. 1894. *An introduction to comparative psychology*. London: Scott.

Morgan, M. J., M. D. Fitch, J. G. Holman, and S. E. Lea. 1976. Pigeons learn the concept of an "A." *Perception*, 5, 57–66.

Muncer, M. J. 1983. "Conservations" with a chimpanzee. *Developmental Psychobiology*, 16, 1–11.

Nagell, K., R. S. Olguin, and M. Tomasello. 1993. Processes of social learning in the tool use of chimpanzees *(Pan troglodytes)* and human children *(Homo sapiens)*. *Journal of Comparative Psychology* 107, 174–186.

Natale, F., F. Antinucci, and P. Poti. 1986. Stage 6 object concept in nonhuman primate cognition: a comparison between gorilla *(G. gorilla gorilla)* and Japanese macaque *(M. fuscata)*. *Journal of Comparative Psychology*, 100, 335–339.

Neisser, U. 1967. *Cognitive psychology*. New York: Appleton-Century-Crofts.

Neiworth, J. J. 1992. Cognitive aspects of movement estimation: A test of imagery in animals. In W. K. Honig and J. G. Fetterman, eds., *Cognitive aspects of stimulus control*, pp. 323–346. Hillsdale, NJ: Lawrence Erlbaum Associates.

Neiworth, J. J., and M. E. Rilling. 1987. A method for studying imagery in animals. *Journal of Experimental Psychology: Animal Behavior Processes*, 13, 203–214.

Neuman, C. J., and S. D. Hill. 1978. Self-recognition and stimulus preference in autistic children. *Developmental Psychobiology*, 12, 85–86.

Nishida, T. 1987. Local traditions and cultural transmissions. In B. B. Smuts, D. L. Cheney, R. M. Seyfarth, R. W. Wrangham, and T. T. Strushaker, eds., *Primate societies*, pp. 462—474. Chicago: University of Chicago Press.

Oden, D. L., R. K. R. Thompson, and D. Premack. 1990. Infant chimpanzees *(Pan troglodytes)* spontaneously perceive both concrete and abstract same/different relations. *Child Development*, 61, 621–631.

Olton, D. S. 1977. Spatial memory. *Scientific American*, June, 82–98.

—— 1978. Characteristics of spatial memory. In S. H. Hulse, H. Fowler and W. K. Honig, eds., *Cognitive processes in animal behavior*, pp. 341–373. Hillsdale, NJ: Lawrence Erlbaum Associates.

Olton, D. S., and R. J. Samuelson. 1976. Remembrance of places passed: Spatial memory in rats. *Journal of Experimental Psychology: Animal Behavior Processes*, 2, 97–116.

Owings, D. H., and R. G. Coss. 1977. Snake mobbing by California ground squirrels: Adaptive variation and ontogeny. *Behaviour*, 62, 50–69.

Owings, D. H., and R. A. Virginia. 1978. Alarm calls of California ground squirrels *(Spermophilus beecheyi)*. *Zeitschrift für Tierpsychologie*, 46, 58–70.

Palameta, B., and L. Lefebvre. 1985. The social transmission of a food-finding technique in pigeons: What is learned? *Animal Behaviour*, 33, 892–896.

Papandropoulou, I., and H. Sinclair. 1974. What is a word? Experimental study of children's ideas on grammar. *Human Development*, 77, 241–258.

Papousek, H., and M. Papousek. 1987. Intuitive parenting: A dialectic counterpart to the infant's integrative competence. In J. D. Osofsky, ed., *Handbook of infant development*, 2nd ed., pp. 669–720. New York: Wiley.

Parker, S. T. 1977. Piaget's sensorimotor series in an infant macaque: A model for comparing unstereotyped behavior and intelligence in human and nonhuman primates. In S. Chevalier-Skolnikoff and F. E. Poirier, eds., *Primate biosocial development: Biological, social, and ecological determinants*, pp. 43–112. New York: Garland.

—— 1990. Origins of comparative developmental evolutionary studies of primate mental abilities. In S. T. Parker and K. R. Gibson, eds., *"Language" and intelligence in monkeys and apes: Comparative developmental perspectives*, pp. 3–64. New York: Cambridge University Press.

—— 1991. A developmental approach to the origins of self-recognition in great apes. *Human Evolution*, 6, 435–449.

Parker, S. T., and K. R. Gibson. 1977. Object manipulation, tool use, and sensorimotor intelligence as feeding adaptations in cebus monkeys and great apes. *Journal of Human Evolution*, 6, 623–641.

—— 1979. A developmental model for the evolution of language and intelligence in early hominids. *The Behavioral and Brain Sciences*, 2, 367–408.

——, eds. 1990. *"Language" and intelligence in monkeys and apes: Comparative developmental perspectives*. New York: Cambridge University Press.

Parker, S. T., and C. Milbrath. 1994. Contributions of imitation and role-playing games to the construction of self in primates. In S. T. Parker, R. W. Mitchell, and M. L. Boccia, eds., *Self-awareness in animals and humans: Developmental perspectives*, pp. 108–128. New York: Cambridge University Press.

Parker, S. T., R. W. Mitchell, and M. L. Boccia, eds. 1994. *Self-awareness in animals and humans: Developmental perspectives.* New York: Cambridge University Press.

Patterson, F. G. 1978. The gestures of a gorilla: Language acquisition in another pongid. *Brain and Language,* 5, 72–97.

Patterson, F. G., and E. Linden. 1981. *The education of Koko.* New York: Holt, Rinehart and Winston.

Patterson, F. G., and Cohn, R. H. 1994. Self-recognition and self-awareness in lowland gorillas. In S. T. Parker, R. W. Mitchell, and M. L. Boccia, eds., *Self-awareness in animals and humans: Developmental perspectives,* pp. 273–290. New York: Cambridge University Press.

Pearce, J. M. 1987. *An introduction to animal cognition.* Hillsdale, NJ: Lawrence Erlbaum Associates.

Pepperberg, I. M. 1994. Numerical competence in an African Grey Parrot *(Psittacus erithacus). Journal of Comparative Psychology,* 108, 36–44.

Pepperberg, I. M., and F. A. Kozak. 1986. Object permanence in the African Grey Parrot *(Psittacus erithacus). Animal Learning and Behavior,* 14, 322–330.

Pepperberg, I. M., and M. S. Funk. 1990. Object permanence in four species of psittacine birds: An African grey parrot *(Psittacus erithacus),* an Illiger mini macaw *(Ara maracana),* a parakeet *(Melopsittacus undulatus),* and a cockatiel *(Nymphicus hollandicus). Animal Learning and Behavior,* 18, 97–108.

Pepperberg, I. M., S. E. Garcia, E. C. Jackson, and S. Marconi. 1995. Mirror use by African grey parrots *(Psittacus erithacus). Journal of Comparative Psychology,* 109, 182–195.

Pereira, M., and J. M. Macedonia. 1991. Ringtailed lemur antipredator calls denote predator class, nor response urgency. *Animal Behaviour,* 41, 543–544.

Petit, O., and B. Thierry. 1993. Use of stones in a captive group of Guinea baboons *(Papio papio). Folia Primatologica,* 61, 160–164.

Piaget, J. 1950. *The psychology of intelligence.* London: Routledge and Kegan Paul.

—— 1952a. *The origins of intelligence in children.* New York: Norton and Co., Inc.

—— 1952b. *The child conception of number.* London: Routledge and Kegan Paul.

—— 1954. *The construction of reality in the child.* New York: Ballantine. [Originally published as *La Construction du Réel chez l'Enfant* (Paris: Delachaux et Niestlé, 1937).]

—— 1962. *Play, dreams, and imitation in childhood.* New York: Norton and Company.

—— 1970. *Genetic epistemology.* New York: Columbia University Press.

—— 1971. *Biology and knowledge.* Chicago: University of Chicago Press.

Piaget, J., and B. Inhelder. 1969. *The psychology of the child.* New York: Basic Books.

Plooij, F. X. 1978. Some basic traits of language in wild chimpanzees? In A. Lock, ed., *Action, gesture and symbol*, pp. 111–131. New York: Academic Press.

Posner, M. I. 1978. *Chronometric exploration of mind.* Hillsdale, NJ: Lawrence Erlbaum Associates.

Poti, P., and G. Spinozzi. 1994. Early sensorimotor development in chimpanzees *(Pan troglodytes). Journal of Comparative Psychology,* 108, 93–103.

Poucet, B., N. Chapuis, M. Durup, and C. Thinus-Blanc. 1986. A study of exploratory behavior as an index of spatial knowledge in hamsters. *Animal Learning and Behavior,* 14, 93–100.

Povinelli, D. J. 1989. Failure to find self-recognition in Asian elephants *(Elephas maximus)* in contrast to their use of mirror cues to discover hidden food. *Journal of Comparative Psychology,* 103, 122–131.

——— 1993. Reconstructing the evolution of mind. *American Psychologist,* 48, 493–509.

Povinelli, D. J., K. E. Nelson, and S. T. Boysen. 1990. Inferences about guessing and knowing by chimpanzees *(Pan troglodytes). Journal of Comparative Psychology,* 104, 203–210.

Povinelli, D. J., K. A. Parks, and M. A. Novak. 1991. Do rhesus monkeys *(Macaca mulatta)* attribute knowledge and ignorance to others? *Journal of Comparative Psychology,* 105, 318–325.

Povinelli, D. J., and S. De Blois. 1992. Young children's *(Homo sapiens)* understanding of knowledge formation in themselves and others. *Journal of Comparative Psychology,* 106, 228–238.

Povinelli, D. J., A. B. Rulf, K. R. Landau, and D. T. Bierschwale 1993. Self-recognition in chimpanzees *(Pan troglodytes):* Distribution, ontogeny, and patterns of emergence. *Journal of Comparative Psychology,* 107, 347–372.

Prato Previde, E., M. Colombetti, M. Poli, and E. C. Spada. 1992. The mind of organisms: Some issues about animal cognition. *International Journal of Comparative Psychology,* 6, 87–100.

Premack, D. 1971. Language in chimpanzees? *Science,* 172, 808–822.

——— 1972. Teaching language to an ape. *Scientific American,* 227, 92–99.

——— 1976. *Intelligence in apes and man.* Hillsdale, NJ: Lawrence Erlbaum Associates.

——— 1983. The codes of man and beast. *The Behavioral and Brain Sciences,* 6, 125–167.

——— 1985. "Gavagai!" or the future of the animal language controversy. *Cognition,* 19, 207–296.

Premack, D., and G. Woodruff. 1978. Does the chimpanzee have a theory of mind? *The Behavioral and Brain Sciences,* 3, 615–636.

Quiatt, D., and V. Reynolds. 1993. *Primate behaviour: Information, social knowledge and the evolution of culture.* New York: Cambridge University Press.

Redshaw, M. 1978. Cognitive development in human and gorilla infants. *Journal of Human Evolution,* 7, 133–141.

Richards, R. J. 1987. *Darwin and the emergence of evolutionary theories of mind and behavior.* Chicago: University of Chicago Press.

Richelle, M., and H. Lejeune. 1980. *Time in animal behavior.* Oxford: Pergamon Press.

Riley, D. A., and C. M. Langley. 1993. The logic of species comparisons. *Psychological Science,* 4, 185–189.

Rilling, M. E., and J. J. Neiworth. 1987. Theoretical and methodological considerations for the study of imagery in animals. *Learning and Motivation,* 18, 57–79.

Ristau, C. A. 1986. Do animals think? In R. J. Hoage and L. Goldman, eds., *Animal intelligence,* pp. 165–185. Washington: Smithsonian Institution Press.

—— 1991a. Aspects of the cognitive ethology of an injury-feigning bird, the piping plover. In C. A. Ristau, ed., *Cognitive ethology: The minds of other animals,* pp. 91–126. Hillsdale: Lawrence Erlbaum Associates.

—— 1991b. Cognitive ethology: An overview. In C. A. Ristau, ed., *Cognitive ethology: The minds of other animals,* pp. 291–313. Hillsdale: Lawrence Erlbaum Associates.

Roberts, W. A., and P. J. Kraemer. 1981. Recognition memory for lists of visual stimuli in monkeys and humans. *Animal Learning and Behavior,* 9, 587–594.

Roberts, W. A., and D. S. Mazmanian. 1988. Concept learning at different levels of abstraction by pigeons, monkeys, and people. *Journal of Experimental Psychology: Animal Behavior Processes,* 14, 247–260.

Roberts, W. A., and M. T. Phelps. 1994. Transitive inference in rats: A test of the spatial coding hypothesis. *Psychological Science,* 5, 368–373.

Roberts, W. A., and W. E. Smythe. 1979. Memory for lists of spatial events in the rat. *Learning and Motivation,* 10, 313–336.

Roitblat, H. L. 1982. The meaning of representation in animal memory. *The Behavioral and Brain Sciences,* 5, 353–406.

—— 1987. *Introduction to comparative cognition.* New York: W. H. Freeman and Company.

Roitblat, H. L., and L. Von Fersen. 1992. Comparative cognition: Representations and processes in learning and memory. *Annual Review of Psychology,* 43, 671–710.

Romanes, G. 1882. *Animal intelligence.* New York: Appleton Co.

Rosch, E. 1978. Principles of categorization. In E. Rosch and B. B. Lloyd, eds., *Cognition and categorization,* pp. 28–49. Hillsdale, NJ: Lawrence Erlbaum Associates.

Rozin, P. 1976. The evolution of intelligence and access to the cognitive unconscious. In J. M. Sprague and A. N. Epstein, eds., *Progress in psychobiology and physiological psychology,* pp. 245–280. New York: Academic Press.

Rumbaugh, D. M., ed. 1977. *Language learning by a chimpanzee: The LANA project.* New York: Academic Press.

Rumbaugh, D. M., and J. L. Pate. 1984. The evolution of cognition in primates: A comparative perspective. In H. L. Roitblat, T. G. Bever, and H. S. Terrace, eds., *Animal cognition*, pp. 569–587. Hillsdale, NJ: Lawrence Erlbaum Associates.

Rumbaugh, D. M., E. S. Savage-Rumbaugh, and M. Hegel. 1987. Summation in the chimpanzee. *Journal of Experimental Psychology: Animal Behavior Processes*, 13, 107–115.

Rumbaugh, D. M., W. K. Richardson, D. A. Washburn, E. S. Savage-Rumbaugh, and W. D. Hopkins. 1989. Rhesus monkeys *(Macaca mulatta)*, video tasks, and implications for stimulus-response spatial contiguity. *Journal of Comparative Psychology*, 103, 32–38.

Rumbaugh, D. M., E. S. Savage-Rumbaugh, and R. A. Sevcik. 1994. Biobehavioral roots of language. In R. W. Wrangham, W. C. Mc Grew, F. B. M. de Waal and P. G. Heltne, eds., *Chimpanzee cultures*, pp. 319–354. Cambridge, MA: Harvard University Press.

Russon, A. E. 1990. The development of peer social interaction in infant chimpanzees: Comparative social, Piagetian, and brain perspectives. In S. T. Parker and K. R. Gibson, eds., *"Language" and intelligence in monkeys and apes: Comparative developmental perspectives*, pp. 379–419. New York: Cambridge University Press.

Russon, A. E., and B. M. F. Galdikas. 1993. Imitation in free-ranging rehabilitant orangutans *(Pongo pygmaeus)*. *Journal of Comparative Psychology*, 107, 147–161.

Ryan, C. M. E., and S. E. G. Lea. 1990. Pattern recognition, updating, and filial imprinting in the domestic chicken *(Gallus gallus)*. In M. L. Commons, R. S. Herrnstein, S. M. Kosslyn, and P. B. Mumford, eds., *Quantitative Analyses of Behaviour*, vol. 8, pp. 89–110. London: Lawrence Erlbaum Associates.

Sands, S. F., and A. A. Wright. 1980. Serial probe recognition performance by a rhesus and a human with 10- and 20-item lists. *Journal of Experimental Psychology: Animal Behavior Processes*, 6, 386–396.

Santiago, H. C., and A. A. Wright. 1984. Pigeon memory: Same/different concept learning, serial probe recognition acquisition, and probe delay effects on the serial-position function. *Journal of Experimental Psychology: Animal Behavior Processes*, 10, 498–512.

Savage-Rumbaugh, E. S. 1986. *Ape language: From conditioned response to symbol*. Oxford: Oxford University Press.

Savage-Rumbaugh, E. S., D. M. Rumbaugh, S. T. Smith, and J. Lawson. 1980. Reference: The linguistic essential. *Science*, 210, 921–925.

Savage-Rumbaugh, E. S., D. M. Rumbaugh, and K. McDonald. 1985. Language learning in two species of apes. *Neuroscience and Biobehavioral Reviews*, 9, 653–665.

Savage-Rumbaugh, E. S., K. McDonald, R. A. Sevcik, W. D. Hopkins, and E. Rupert. 1986. Spontaneous symbol acquisition and communicative use by pygmy chimpanzees *(Pan paniscus)*. *Journal of Experimental Psychology: General*, 115, 211–235.

Savage-Rumbaugh, E. S., J. Murphy, R. A. Sevcik, K. E. Brakke, S. L. Williams, and D. M. Rumbaugh. 1993. Language comprehension in ape and child. *Monographs of the Society for Research in Child Development,* 58 (3–4).

Schaffer, H. R. 1984. *The child's entry into a social world.* London: Academic Press.

Schino, G., G. Spinozzi, and L. Berlinguer. 1990. Object concept and mental representation in *Cebus apella* and *Macaca fascicularis. Primates,* 31, 537–544.

Schrier, A. M., R. Angarella, and M. Povar. 1984. Studies of concept formation by stumptail monkeys: Concepts monkeys, humans and letter A. *Journal of Experimental Psychology: Animal Behavior Processes,* 10, 564–584.

Schull, J., and J. D. Smith. 1992. Knowing thyself, knowing the other: They're not the same. *The Behavioral and Brain Sciences,* 15, 166–167.

Schusterman, R. J., and R. Gisiner. 1988. Artificial language comprehension in dolphins and sea lions: The essential cognitive skills. *The Psychological Record,* 38, 311–348.

Schusterman, R. J., and K. Krieger. 1984. California sea lions are capable of semantic comprehension. *The Psychological Record,* 34, 3–23.

Schusterman, R. J., J. A. Thomas, and F. G. Wood, eds. 1986. *Dolphin cognition and behavior: A comparative approach.* Hillsdale, NJ: Lawrence Erlbaum Associates.

Seligman, M. E. P. 1970. On the generality of the laws of learning. *Psychological Review,* 77, 406–418.

Seyfarth, R. M., and D. L. Cheney. 1984. Grooming, alliances and reciprocal altruism in vervet monkeys. *Nature,* 308, 541–543.

Seyfarth, R. M., D. L. Cheney, and P. Marler. 1980. Monkey responses to three different alarm calls: Evidence of predator classification and semantic communication. *Science,* 210, 801–803.

Shepard, R. N. 1982. *Mental images and their transformations.* Cambridge, MA: MIT Press.

Shepard, R. N., and J. Metzler. 1971. Mental rotation of three-dimensional objects. *Science,* 171, 701–703.

Sherry, D. F. 1984. What food-storing birds remember. *Canadian Journal of Psychology,* 38, 304–321.

Shettleworth, S. J. 1983. Memory in food-storing birds. *Scientific American,* 248, 102–110.

———— 1993. Where is the comparison in comparative cognition? Alternative research programs. *Psychological Science,* 4, 179–184.

Shettleworth, S. J., and J. R. Krebs. 1982. How marsh tits find their hoards: The roles of site preference and spatial memory. *Journal of Experimental Psychology: Animal Behavior Processes* 8, 354–375.

Smith, J. D., J. Schull, D. A. Washburn, and W. E. Shields. 1994. Uncertainty monitoring in the rhesus monkey *(Macaca mulatta).* In J. R. Anderson, J.-J. Roeder, B. Thierry, and N. Herrenschmidt, eds., *Current primatology,* vol. 3:

Behavioral neuroscience, physiology and reproduction, pp. 101–110. Strasbourg: Université Louis Pasteur.

Snowdon, C. T. 1990. Language capacities of nonhuman animals. *Yearbook of Physical Anthropology,* 33, 215–243.

Snowdon, C. T., and J. A. French. 1979. Ontogeny does not always recapitulate phylogeny: Open peer commentary to "A developmental model for the evolution of language and intelligence in early hominids." *The Behavioral and Brain Sciences,* 2, 397–398.

Spence, K. W., and R. O. Lippitt. 1940. "Latent" learning of a simple maze problem in the rat. *Psychological Bulletin,* 37, 429.

Spinozzi, G., and F. Natale. 1986. The interaction between prehension and locomotion in macaque, gorilla and child cognitive development. In J. G. Else and P. C. Lee, eds., *Primate ontogeny, cognition and social behaviour,* pp. 155–160. New York: Cambridge University Press.

Spinozzi, G., and P. Poti. 1993. Piagetian stage 5 in two infant chimpanzees *(Pan troglodytes):* The development of permanence of objects and the spatialization of causality. *International Journal of Primatology,* 14, 905–917.

Straub, R. O., and H. S. Terrace. 1981. Generalization of serial learning in the pigeon. *Animal Learning and Behavior,* 9, 454–468.

Strushaker, T. T. 1967. Auditory communication among vervet monkeys *(Cercopithecus aethiops).* In S. A. Altman, ed., *Social communication among primates,* pp. 281–324. Chicago: Chicago University Press.

Struhsaker, T. T., and P. Hunkeler. 1971. Evidence of tool-using by chimpanzees in the Ivory Coast. *Primates,* 15, 212–219.

Suarez, S. D., and G. G. Gallup. 1981. Self-recognition in chimpanzees and orangutans, but not gorillas. *Journal of Human Evolution,* 10, 175–188.

Suboski, M. D., D. Muir, and D. Hall. 1993. Social learning in invertebrates. *Science,* 259, 1628–1629.

Sugiyama, Y., and J. Koman. 1979. Tool-using and tool-making behavior in wild chimpanzees at Bossou, Guinea. *Primates,* 20, 513–524.

Suzuki, S., G. Augerinos, and A. H. Black. 1980. Stimulus control of spatial behavior in the eight-arm maze in rats. *Learning and Motivation,* 11, 1–18.

Swartz, K. B., and S. Evans. 1991. Not all chimpanzees *(Pan troglodytes)* show self-recognition. *Primates,* 32, 483–496.

Tayler, C. K., and G. S. Saayman. 1973. Imitative behaviour by Indian Ocean bottlenose dolphins *(Tursiops aduncus)* in captivity. *Behaviour,* 44, 286–298.

Terrace, H. S. 1979. *Nim: A chimpanzee who learned Sign Language.* New York: Knopf.

—— 1984. Animal cognition. In H. L. Roitblat, T. G. Bever, and H. S. Terrace, eds., *Animal cognition,* pp. 7–28. Hillsdale, NJ: Lawrence Erlbaum Associates.

—— 1985. In the beginning was the "name." *The American Psychologist,* 40, 1011–1028.

———— 1993. The phylogeny and ontogeny of serial memory: List learning by pigeons and monkeys. *Psychological Science,* 4, 162–169.

Terrace, H. S., L. A. Petitto, S. J. Sanders, and T. G. Bever. 1979. Can an ape create a sentence? *Science,* 200, 891–902.

Thinus-Blanc, C., and P. Scardigli. 1981. Object permanence in the golden hamster. *Perceptual and Motor Skills,* 53, 1010.

Thinus-Blanc, C., B. Poucet, and N. Chapuis. 1982. Object permanence in cats: Analysis in locomotor space. *Behavioural Processes,* 7, 81–86.

Thinus-Blanc, C., L. Bouzouba, K. Chaix, N. Chapuis, M. Durup, and B. Poucet. 1987. A study of spatial parameters encoded during exploration in hamsters. *Journal of Experimental Psychology: Animal Behavior Processes,* 13, 418–427.

Thomas, R. K. 1980. Evolution of intelligence: An approach to its assessment. *Brain Behavior and Evolution,* 17, 454–472.

Thomas, R. K., and L. Peay. 1976. Length judgments by squirrel monkeys: Evidence for conservation? *Developmental Psychology,* 12, 349–352.

Thomas, R. K., and E. L. Walden. 1985. The assessment of cognitive development in human and non-human primates. In E. S. Watts, ed., *Nonhuman primate models for human growth and development,* pp. 187–215. New York: Alan R. Liss, Inc.

Thompson, R. K. R. 1995. Natural and relational concepts in animals. In H. Roitblat and J. A. Meyer, eds., *Comparative approaches to cognitive science,* pp. 175–224. Cambridge, MA: Bradford Books, MIT Press.

Thompson, R. L., and S. L. Boatright-Horowitz. 1994. The question of mirror-mediated self-recognition in apes and monkeys: Some new results and reservations. In S. T. Parker, R. W. Mitchell, and M. L. Boccia, eds., *Self-awareness in animals and humans: Developmental perspectives,* pp. 330–349. New York: Cambridge University Press.

Thompson, R. K. R., and C. L. Contie. 1994. Further reflections on mirror usage by pigeons: Lessons from Winnie-the-Pooh and Pinocchio too. In S. T. Parker, R. W. Mitchell, and M. L. Boccia, eds., *Self-awareness in animals and humans: Developmental perspectives,* pp. 392–409. New York: Cambridge University Press.

Thompson, R. K. R., and J. Demarest. 1992. Comparative psychology: Last bastion of a compleat functionalism. In D. A. Owens and M. Wagner, eds., *Progress in modern psychology: The legacy of American functionalism,* pp. 55–72. Westport, CT: Praeger.

Thompson, R. K. R., and L. M. Herman. 1977. Memory for lists of sounds by the bottle-nosed dolphin: Convergence of memory processes with humans. *Science,* 195, 501–503.

Thorndike, E. L. 1898. Animal intelligence: An experimental study of the associative processes in animals. *Psychological Review, Monographs Supp.* 28, 1–109.

———— 1911. *Animal intelligence: Experimental studies.* New York: Macmillan.

Thorpe, W. H. 1956. *Learning and instinct in animals.* London: Methuen.

——— 1972. The comparison of vocal communication in primates and in man. In R. A. Hinde, ed., *Non-verbal communication,* pp. 27–47. Cambridge, MA: Harvard University Press.

——— 1979. *The origins and rise of ethology.* London: Heinemann.

Timberlake, W., and W. White. 1990. Winning isn't everything: Rats need only food deprivation and not food reward to efficiently traverse a radial arm maze. *Learning and Motivation,* 21, 153–163.

Tinbergen, N. 1963. On aims and methods of ethology. *Zeitschrift für Tierpsychologie,* 20, 410–433.

Tinklepaugh, O. L. 1928. An experimental study of representative factors in monkeys. *Journal of Comparative Psychology,* 8, 197–236.

——— 1932. Multiple delayed reaction with chimpanzees and monkeys. *Journal of Comparative Psychology,* 13, 207–243.

Todrin, S. C., and D. S. Blough. 1983. The discrimination of mirror-image forms by pigeons. *Perception and Psychophysics,* 34, 397–402.

Tokida, E., I. Tanaka, H. Takefushi, and T. Hagiwara. 1994. Tool-using in Japanese macaques: Use of stones to obtain fruit from a pipe. *Animal Behaviour,* 47, 1023–1030.

Tolman, E. C. 1932. *Purposive behavior in animals and men.* New York: Appleton-Century-Crofts.

——— 1948. Cognitive map in rats and men. *Psychological Review,* 55, 189–209.

Tomasello, M. 1990. Cultural transmission in the tool use and communicatory signaling of chimpanzees? In S. T. Parker and K. R. Gibson, eds., *"Language" and intelligence in monkeys and apes: Comparative developmental perspectives,* pp. 275–311. New York: Cambridge University Press.

Tomasello, M., M. Davis-Dasilva, L. Camak, and K. A. Bard. 1987. Observational learning of tool-use by young chimpanzees. *Human Evolution,* 2, 175–83.

Tomasello, M., A. C. Kruger, and H. H. Ratner. 1993. Cultural learning. *The Behavioral and Brain Sciences,* 16, 495–552.

Tomasello, M., S. Savage-Rumbaugh, and A. C. Kruger. 1993. Imitative learning of actions on objects by children, chimpanzees, and enculturated chimpanzees. *Child Development,* 64, 1688–1705.

Torigoe, T. 1985. Comparison of object manipulation among 74 species of non-human primates. *Primates,* 26, 182–194.

——— 1986. Object manipulation in primates: A comparative psychological approach to human behavior. *Hiroshima Forum for Psychology,* 11, 89–99.

Triana, E., and R. Pasnak. 1981. Object permanence in cats and dogs. *Animal Learning and Behavior,* 9, 135–139.

Ullman, S. 1978. Mental representations and mental experiences. *The Behavioral and Brain Sciences,* 4, 605–606.

Van Beusekom, G. 1948. Some experiments on the optical orientation in *Philanthus triangulum* Fabr. *Behaviour,* 1, 195–225.

Vallortigara, G., M. Zanforlin, and G. Pasti. 1990. Geometric modules in animals' spatial representations: A test with chicks *(Gallus gallus domesticus)*. *Journal of Comparative Psychology*, 104, 248–254.

Vauclair, J. 1982. Sensorimotor intelligence in human and non-human primates. *Journal of Human Evolution*, 11, 257–264.

——— 1984. Phylogenetic approach to object manipulation in human and ape infants. *Human Development*, 27, 321–328.

——— 1987. A comparative approach to cognitive mapping. In P. Ellen and C. Thinus-Blanc, eds., *Cognitive processes and spatial orientation in animal and man*, vol. 1: *Experimental psychology and ethology*, pp. 89–96. Dordrecht: Martinus Nijhoff.

——— 1989. Effects of different types of visual information on the baboon's spatial representation and memory. *Bulletin of the Psychonomic Society*, 276, 501.

——— 1990a. Processus cognitifs élaborés: Etude des représentations mentales chez le babouin. In J.-J. Roeder and J. R. Anderson, eds., *Primates: Recherches actuelles*, pp. 170–180. Paris: Masson.

——— 1990b. Primate cognition: From representation to language. In S. T. Parker and K. R. Gibson, eds., *"Language" and intelligence in monkeys and apes: Comparative developmental perspectives*, pp. 312–329. New York: Cambridge University Press.

Vauclair, J., and J. R. Anderson. 1994. Object manipulation, tool use, and the social context in human and nonhuman primates. *Techniques and Cultures*, B24, 121–126.

Vauclair, J., and K. A. Bard. 1983. Development of manipulations with objects in ape and human infants. *Journal of Human Evolution*, 12, 631–645.

Vauclair, J., and J. Fagot. 1993. Manual and hemispheric specialization in the manipulation of a joystick by baboons. *Behavioral Neuroscience*, 107, 210–214.

Vauclair, J., J. Fagot, and W. D. Hopkins. 1993. Rotation of mental images in baboons when the visual input is directed to the left cerebral hemisphere. *Psychological Science*, 4, 99–103.

Vauclair, J., and J. Fagot. 1996. Extent and limits of categorization in baboons *(Papio papio)*: Assessment with identity and arbitrary matching-to-sample tasks. (Submitted.)

Vauclair, J., and J. M. Vidal. 1994. Discontinuities in the mind between animals and humans. Paper presented at the conference "Cognition and Evolution," Berder Island, March 10.

Vaughter, R. M., W. Smotherman, and J. M. Ordy. 1972. Development of object permanence in the infant squirrel monkey. *Developmental Psychology*, 7, 34–38.

Visalberghi, E. 1986. Aspects of space representation in an infant gorilla. In D. M. Taub and F. A. King, eds., *Current perspectives in primate social dynamics*, pp. 445–452. New York: Van Nostrand Reinhold Company.

—— 1990. Tools use in *Cebus. Folia Primatologica,* 54, 146–154.

—— 1992. Is lack of understanding of cause-effect relationships a suitable basis for interpreting monkeys' failures in attribution? *The Behavioral and Brain Sciences,* 15, 169–170.

Visalberghi, E., and D. Fragaszy. 1990. Do monkeys ape? In S. T. Parker and K. R. Gibson, eds., *"Language" and intelligence in monkeys and apes: Comparative developmental perspectives,* pp. 247–273. New York: Cambridge University Press.

Visalberghi, E., and L. Trinca. 1989. Tool use in capuchin monkeys, or distinguishing between performing and understanding. *Primates,* 30, 511–521.

Von Fersen, L., C. D. L. Wynne, J. D. Delius, and J. E. R. Staddon. 1991. Transitive inference formation in pigeons. *Journal of Experimental Psychology: Animal Behavior Processes,* 17, 334–341.

Von Frisch, K. 1950. *Bees, their vision, chemical senses and language.* Oxford: Oxford University Press.

—— 1967. *The dance language and orientation of bees.* Cambridge, MA: Harvard University Press.

Von Glaserfeld, E. 1976. Linguistic communication: Theory and definition. In D. M. Rumbaugh, ed., *The LANA Project,* pp. 55–71. New York: Academic Press.

—— 1977. The development of language as purposive behavior. In S. R. Harnad, H. D. Steklis, and J. Lancaster, eds., *Origins and evolution of language and speech,* pp. 212–226. New York: New York Academy of Sciences.

Vygotsky, L. S. 1962. *Thought and language.* Cambridge, MA: MIT. Press.

Walker, S. 1983. *Animal thought.* London: Routledge and Kegan.

Wallman, J. 1992. *Aping language.* New York: Cambridge University Press.

Warren, J. M. 1969. Discrimination of mirror-images by cats. *Journal of Comparative and Physiological Psychology,* 69, 9–11.

Washburn, M. F. 1908. *The animal mind.* New York: Macmillan.

Wasserman, E. A. 1981. Comparative psychology returns: A review of Hulse, Fowler, and Honig's "Cognitive Processes in Animal Behavior." *Journal of the Experimental Analysis of Behavior,* 35, 243–257.

—— 1993. Comparative cognition: Beginning the second century of the study of animal intelligence. *Psychological Bulletin,* 113, 211–228.

Watson, J. B. 1919. *Psychology from the standpoint of a behaviorist.* Philadelphia: Lippincott.

Wehner, R., and R. Menzel. 1990. Do insects have cognitive maps? *Annual Review of Neuroscience,* 13, 403–414.

Wellman, H. M. 1990. *The child's theory of mind.* Cambridge, MA: MIT Press.

Westergaard, G. C., and D. M. Fragaszy. 1987. The manufacture and use of tools by capuchin monkeys *(Cebus apella). Journal of Comparative Psychology,* 101, 159–168.

Whiten, A., and R. W. Byrne. 1988. Tactical deception in primates. *The Behavioral and Brain Sciences*, 11, 233–273.

Wiley, R. H., and S. A. Hartnett. 1980. Mechanisms of spacing in groups of juncos: Measurement of behavioral tendencies in social situations. *Animal Behaviour*, 28, 1005–1016.

Wiltschko, W., and R. Wiltschko. 1987. Cognitive maps and navigation in homing pigeons. In P. Ellen and C. Thinus-Blanc, eds., *Cognitive processes and spatial orientation in animal and man*, vol. 1: *Experimental psychology and ethology*, pp. 201–216. Dordrecht: Martinus Nijhoff.

Winnicott, D. W. 1971. *Playing and reality*. Basic Books: New York.

Wolfe, J. B. 1936. Effectiveness of token rewards for chimpanzees. *Comparative Psychology Monographs*, 12, 1–72.

Wood, S., K. M. Moriarty, B. T. Gardner, and R. A. Gardner. 1980. Object permanence in child and chimpanzee. *Animal Learning and Behavior*, 8, 3–9.

Woodruff, G., and D. Premack. 1979. Intentional communication in the chimpanzee: The development of deception. *Cognition*, 7, 333–362.

——— 1981. Primitive mathematical concepts in the chimpanzee: Proportionality and numerosity. *Nature*, 293, 568–570.

Woodruff, G., D. Premack, and K. Kennel. 1978. Conservation of liquid and solid quantity by the chimpanzee. *Science*, 202, 991–994.

Wright, A. A. 1994. Primacy effects in animal memory and human nonverbal memory. *Animal Learning and Behavior*, 22, 219–223.

Wright, A. A., H. C. Santiago, S. F. Sands, and D. F. Kendrick. 1985. Memory processing of serial lists by pigeons, monkeys and people. *Science*, 229, 287–289.

Wynne, C. D. L. 1995. Reinforcement account for transitive inference performance. *Animal Learning and Behavior*, 23, 207–217.

Yerkes, R. M. 1911. *Introduction to psychology*. New York: Henry Holt.

Yoerg, S. I., and A. C. Kamil. 1991. Integrating cognitive ethology with cognitive psychology. In C. A. Ristau, ed., *Cognitive ethology: The minds of other animals*, pp. 271–290. Hillsdale, NJ: Lawrence Erlbaum Associates.

Young, J. Z. 1991. Computation in the learning systems of cephalopods. *Biological Bulletin*, 180, 200–208.

Zayan, R. ed. 1994. Individual and social recognition. Special issue of *Behavioural Processes*, 33 (1–2).

Zentall, T. R., and D. E. Hogan. 1976. Pigeons can learn identity or difference, or both. *Science*, 191, 408–409.

Zivin, G., ed. 1979. *Development of self-regulation through speech*. New York: Wiley.

Index

Abstract relations, 12, 14
Abstract representations, 14–15
Accommodation, 29, 31, 32
Acquisition, 31–35, 51, 126, 150–151
Adaptation, 1–2, 5, 29–31, 161, 167–168;
  adaptive value, 38, 171
Affiliation, 90, 92, 132
Alarm calls, 93, 103–104, 117, 140–141
Alliances, 84, 95
Altruistic acts, 95
Analogy: in evolution, 166
Analogy in reasoning, 47, 92
Animal cognition vs. human cognition, 29–
  31, 50, 130, 155, 171–172
Anthropomorphism, 50–51, 95
Anticipation, 40, 98, 157
Ants, 59, 63, 94–95
Arbitrariness in symbol use, 102, 106, 118–
  119
ASL (American Sign Language), 107
Assimilation, 31–32
Attention, 88, 114, 116, 120–123
Attribution of mental states, 125–126, 132–
  133, 135, 137, 139, 141, 147–149, 151–152

Baboons, 24–27, 58, 69–70, 77–80, 93–94,
  125, 135, 150, 153
Bees, 63–65, 68, 76, 102–103, 156–157, 165
Behavioral ecology, 165
Behaviorism, 3–4; and mental processes, 155–
  156

Beliefs. See Attribution of mental states
Bidirectional control procedure, 128
Bonobos (pygmy chimpanzees), 113, 115,
  120–121, 169–170
Brain, 6, 81, 106, 168–169; size of, 26, 97,
  108; split-brain, 26
Budgerigars, 128
Bushbabies, 56

Capuchin monkeys, 17, 37, 56–57, 145, 153.
  See also Cebus monkeys
Categorization, 12, 15–16, 35, 160–161, 166
Cats, 3, 15, 24, 31, 35–36, 70, 116
Causality, 33–34, 39–40, 130, 137, 148–149,
  153
Cebus monkeys, 38, 50–51, 167, 170
Cerebral hemispheres, 24, 26
Chickadees, 19, 65
Chicks, 36, 75–76, 103, 133–134
Children, 29–35, 38, 44, 48–49, 51, 62, 115–
  125, 139, 148–149; autistic, 149
Chimpanzees, 14, 38, 40–50, 53–54, 56–62,
  66–69, 76–79, 94, 106–125, 132–133, 135–
  144, 148–154
Church/Gibbon timing model, 80–81
Circular reactions, 32–33, 41, 121–122
Classification of dominance, 90–92
Cognitive ethology, 27, 155–157
Cognitive maps, 5–6, 60, 62–71, 73, 76–80;
  defined, 6
Cognitive sciences, 7, 29, 163, 171

Concept learning. *See* Categorization
Conservation tasks, 42, 48–49
Communication, 2, 86–87, 95, 107, 117–118, 120, 156, 169; defined, 99–102; gestural, 86, 108, 156; referential, 123; symbolic, 113; verbal, 124
Comparisons between species, 3, 60, 68–69, 102, 123, 151, 161–172
Conditioning, 21, 43, 52, 92, 110, 162; aversive, 135; instrumental, 160; Pavlovian, 160
Consciousness, 2, 4, 10, 155–156; defined, 27
Constraints on learning, 163, 165
Control of subjective states, 141, 145–147
Corvids, 44
Counting, 44–47
Crickets, 63
Culture, 35, 131–132

Darwinian model, 1–3, 31, 164, 169
Deceit, 85, 97, 102, 126, 135–137, 148
Deception. *See* Deceit
Delayed matching-to-sample (DMTS) tasks, 19, 24–25
Delayed responses, 4–5
De Saussure, F., 99, 118
Designation, 100, 104, 119
Development. *See* Ontogeny
Discontinuity in cognition, 2, 31, 169
Displacement: as a linguistic feature, 102, 115; the practical group of displacements, 34
Dogs, 4–5, 36, 41, 70
Dolphins, 14, 19, 21, 27, 83, 86, 108, 110, 129
Dualism, 2
Duality of patterning, 102, 106, 124

Eagles, 104
Ecological niche, 163
Ecological validity, 31, 79, 96
Elephants, 55, 143, 153
Enculturation, 151
Ethology, 30, 164–165
Evolution, 96, 165, 172; of language and intelligence, 50, 166, 170; of metacognition, 154; of species, 1, 92
Evolutionary continuity, 1–2, 100, 169
Evolutionary significance, 42, 118
Eye-to-eye contacts, 123, 144–145

Feeding behavior, 80, 99, 161
Finches, 55, 60
Fishes, 86

Fixed-action patterns, 30
Foraging behavior, 64, 96–97, 161, 163, 168
Fractional antedating goal responses, 6

Gestalt psychology, 7
Gestures, 85, 100, 106–110, 114–116, 122, 124, 129–130
Gibbons, 56
Gorillas, 38–40, 56, 58, 68, 83, 108, 121, 123, 142, 156
Grammar, 107, 114–115, 117
Grebes, 159
Gulls, 61

Habit family hierarchies, 6
Habituation procedure, 69
Hamsters, 36, 69
Hierarchy: in object combination, 62; in social relations, 92, 160–161; in tool use, 61
Hippopotamus, 93
Homology, 166

Imagery, 12, 16, 21–22
Imitation, 2, 30, 82, 107, 114, 126–132, 147–151, 153; deferred, 100; symbolic, 130–131
Individual recognition, 86, 89, 165
Infancy, 14, 32, 120, 130, 169
Inference. *See* Inferential reasoning
Inferential reasoning, 35, 42–44, 91–92, 106, 148
Information: dissimulation of, 134–135, 168
Information processing theories, 1, 7–10, 31
Insight behavior, 6, 13
Intelligence: development of, 28–29, 31, 34, 48
Intentionality, 102, 122; defined, 157
Intentions, 98, 125–127, 130, 134, 156–157, 169
Intermodal equivalence, 130, 145
Internal clock, 80–81
Invertebrates, 53, 55, 63, 127, 157

Joint attention, 116, 123, 125
Juncos, 160–161

Language, 2, 14, 21, 26, 34, 49–50, 82, 86, 98–102, 106–107, 110, 114–120, 126–127, 158, 163, 166–170; acquisition of, 116, 123; comprehension, 169–170; declarative function of, 116; defined, 100, 168; main features of, 102, 118, 123–125; production of, 169–170; origins of, 100–101

Langurs, 56
Leaf clipping, 132
Learning: associative, 44; discriminative, 12, 89; latent, 5–6; trial-and-error, 57; observational, 150, 163
Learning sets, 12–13
Lemurs, 50, 56, 104
Leopards, 58, 104
List learning, 16, 18. *See also* Serial learning
Local-specific behaviors, 132
Locomotion, 38–39, 80, 122

Macaques, 12–14, 19, 88–91, 130–132, 139–143, 146
Maps. *See* Cognitive maps
Marmosets, 56, 70, 162
Matching to sample (MTS) tasks, 13–14, 22, 25, 27, 46, 89, 160
Matriline, 90, 95, 132
Maze, 3, 6, 70–71, 74–75. *See also* Radial maze
Means-end relations. *See* Causality
Memory, 7–9, 19–20, 76–80; defined, 9; short-term, 131; spatial, 65, 68, 76, 78, 162, 167; reference, 81; working, 71, 81
Mental chronometry, 1
Mental events, 7
Mental experiences, 27, 148, 155–156
Mental images, 20–21, 24, 34, 118, 158
Mental operations, 31, 35, 158
Mental representations, 27, 57, 130, 155–156
Mental rotation, 20, 23–27
Mentalism, 4, 135
Mirror-guided behaviors, 143–145, 153
Mirror-image discrimination, 22–27
Mirror self-recognition, 141–145, 147–149, 153, 156
Modes of social exchanges, 86, 95, 120, 122
Modes of symbolization, 158
Monocular vision, 78–79
Morgan's canon, 2–3, 152
Morphemes, 192
Motives. *See* Attribution of mental states

Neurosciences, 82, 171
Number concept. *See* Counting
Nutcrackers, 65, 167

Object manipulation, 30, 41–42, 56, 61, 120–121, 123
Object permanence, 5, 31–40, 153
Octopus, 127–128

Olfactory cues, 43, 45, 74
Ontogeny, 30–31, 50, 52, 122, 168
Orangutans, 56, 58, 108, 113, 123, 142, 156

Parrots, 36, 44, 128, 143
Pedagogy, 150, 152
Perception, 78–79; of songs, 165
Perception-knowledge relations, 137–139
Perceptual indices, 158
Perspective taking, 139, 149
Phonemes, 102
Phylogeny, 31, 50–51, 94, 108, 171
Piagetian: stages, 5, 31–35; theory, 28–32
Pigeons, 10, 14–18, 21–24, 26, 42–43, 128, 143–144, 153, 167
Play, 41, 107, 120, 124; contingent, 149, 153; pretend, 124–125; symbolic, 118, 124–125, 153
Playback experiments, 80, 87, 93, 103–104, 106
Plovers, 134
Pointing gestures, 46, 115, 120, 136–138
Prehension, 30, 32
Prevarication. *See* Deceit
Primacy effect. *See* Serial position effects
Principle of parsimony. *See* Morgan's canon
Problem solving, 5–6, 8, 12, 71, 165, 168, 170
Proprioception, 130–131
Protected threat, 93–94
Protoculture. *See* Culture
Psychology: cognitive, 1, 7–8, 53, 157, 171; comparative, 1–3, 5, 27–29, 31, 82, 97, 152, 156, 162, 164, 168–169, 172; developmental, 29, 114, 152

Rabbits, 3
Radial maze, 71, 73, 75, 161. *See also* Maze
Radical arbitrariness, 118–119. *See also* Arbitrariness in symbol use
RAM interpretation (reasoning about mental states), 137, 141, 144
Rank representation, 91, 94, 105, 160
Rats, 6, 42–45, 71–75, 80–81, 128–129, 145, 161–162
Recapitulationism, 50
Recency effect. *See* Serial position effects
Referent, 34, 105, 115, 118–120
Referential abilities, 103, 113, 116, 123
Reflexes, 30–31
Representation: defined, 9, 100; of body, 143, 145, 156; of space, 9, 53–54, 59–60, 62–65,

Representation *(continued)*
68–72, 75–76; of time, 53, 62, 80–81, 166; types of, 157–161
Reptiles, 157
RO interpretation (reasoning about observables), 137, 141
Romanes, G., 2–3
Rules of action, 161–162

Schemes, 29–33, 41
Sea lions, 108–109, 110
Sea otters, 55, 82, 108
Self-awareness, 143–144, 148, 152–153, 156, 169
Self-directed behaviors, 142–143, 145, 149, 153
Self-knowledge. *See* Self-recognition
Self-recognition, 126, 141–143, 148–149, 152–154, 156. *See also* Mirror self-recognition
Semanticity, 101, 103–104, 106, 166
Sensorimotor: abilities, 30, 53, 85; acquisitions, 32–34, 130, 149; stages, 32–34, 149
Serial learning, 12, 16–18
Serial position effects, 18–20, 166
Serial probe recognition (SPR), 19–20
Siamangs, 56
Signifier, 34, 118–119
Simultaneous chaining procedure. *See* Serial learning
Simultaneous matching-to-sample (SMTS) tasks, 90
Social attribution. *See* Attribution of mental states
Social convention, 99–100, 119, 121, 131
Social enhancement. *See* Social learning
Social learning, 82, 127–128, 132, 151; defined, 127
Software, 85, 122
Spatial memory. *See* Memory
Spatial representation. *See* Representation
Spatial orientation, 6, 10, 165
Speech, 82, 102, 106, 115, 170
Squirrel monkeys, 19, 37, 42–43, 50
Squirrels, 55, 104
S-R theory, 6, 8, 30
Stilts, 93
Subjective inference, 2
Subjective mental experiences, 27, 156
Substitute, 4, 46, 99–100, 112, 158–159
Symbolic function, 34, 100
Symbolic matching procedure, 108

Symbolization process, 103, 158
Symbols, 99–100, 111–113, 115–116, 118
Syntax, 101, 117; rules of, 82, 106–107
Synthetic approach. *See* Behavioral ecology

Tamarins, 56
Tarsiers, 56
Teaching, 106, 110, 131, 147–152. *See also* Pedagogy
Temporal and spatial couplings, 160
Temporal generalization, 80
Temporal patterning, 162
Temporal regulations, 82
Temporal representations. *See* Representation
Termites, 53–54, 59, 82
Theory of mind, 98–99, 125–126, 133, 145, 148–149, 152
Tinklepaugh, O. L., 4–5
Tits, 65, 162
Tolman, E. C., 5–6
Tool use, 42, 51, 53–62, 81–83, 122–123, 159, 170; acquisition of, 62, 82, 150–151; defined, 54; functions of, 60–61; manufacture of tools, 55, 59; social tool use, 58, 61, 93–95, 97
Transitional object, 124
Transitive inference. *See* Inferential reasoning
Transitivity, 35, 43, 46
Triadic relationships, 124–125
Tropism, 3
Two-choice discrimination procedure, 89–91

Uncertainty monitoring. *See* Control of subjective states

Vertebrates, 5, 53, 65, 94, 127, 157, 163
Vervet monkeys, 87–88, 93, 95, 104, 117, 125, 160–161
Video image, 45, 78–80, 103, 153
Vocabulary, 107, 117
Vocalizations, 87–88, 100, 103–105, 133–135
Vocal recognition, 87–88

Watson, J. B., 3–4
Wasps, 55, 63, 82
Win/shift strategy, 161–162
Win/stay strategy, 162
Words, 99–100, 102, 106–107, 110–111, 116–119, 169–170
Wolves, 41